A TRUCKER'S TALE

WIT, WISDOM, AND
TRUE STORIES FROM
60 YEARS ON THE ROAD

A TRUCKER'S TALE

▶ ED MILLER

APOLLO
PUBLISHERS

A Trucker's Tale: Wit, Wisdom, and True Stories from 60 Years on the Road

Copyright © 2023 by Ed Miller

All rights reserved. No part of this book may be used or reproduced in any manner whatsoever without the written permission of the publisher, except in the case of brief excerpts in critical reviews or articles. All inquiries should be sent by email to Apollo Publishers at info@apollopublishers.com.

Apollo Publishers books may be purchased for educational, business, or sales promotional use. Special editions may be made available upon request. For details, contact Apollo Publishers at info@apollopublishers.com.

Visit our website at www.apollopublishers.com.

Library of Congress Cataloging-in-Publication Data is available on file.

Print ISBN: 978-1-954641-81-5
Ebook ISBN: 978-1-948062-39-8

Printed in the United States of America.
First printed in hardcover in 2020.

A Trucker's Tale is dedicated to the millions of truckers, both past and present, who have helped America stay strong and free.

CONTENTS

PREFACE

Surely everyone knows that a fishing story grows each time it's told. A minnow morphs into a largemouth bass after just a few beers in a bar full of new faces. You might have heard some awfully tall yarns spun by drivers, maybe at a truck stop lunch counter while you sopped up your eggs with toast and bacon. There's the one driver who stopped along the highway to help a little old lady fix a flat tire. The third time he tells this story, the little old lady becomes a buxom blonde twenty-one-year-old. Perhaps by the fifth retelling, she takes him up on his offer for a ride.

I don't have many tall tales to offer—the stories in this book truly happened to me or to truckers I know. Some names have been changed in a good-faith effort to protect the identities of the bone-headed, dim-witted, and off-kilter, or because I don't want my ass whipped for telling the truth about those of you who might prefer to remain anonymous.

I've been part of the trucking world for sixty years, and I'm damn proud of it. I was born into a trucking family, and as soon as I could talk, I was pestering my dad to ride in his truck. Each time I asked, he would tell me that I could ride with him when I was old enough to climb into the truck without any help. I must have been five or six years old when I climbed onto the running board, the side step, and crawled up into his Mack B61. I'd known the smell of diesel since I was three or four, but the diesel smell from the B61 was unique, and awesome. In later years I would come to associate the smell with a flash of lightning—fierce, quick, and powerful. It burns the nostrils, leaves the tongue bristling, and makes your arm hair stand up. For me the smell conjures feelings of power and brings an adrenaline high. It's a symbol of a journey about to be undertaken.

Several years ago, I was privileged to be the guest speaker at a dinner for the Maryland Motor Truck Associations's annual Truck Driving Championships awards ceremony. The competition dates to 1955, and competitors are tested on their driving and inspection skills, knowledge, and professionalism. Winners qualify to compete in the American Trucking Associations's annual National Truck Driving Championships. Anyway, I began my talk by asking how many of the several hundred truckers in attendance had grown up in trucking families, and the majority of the drivers raised their hands. I asked how many of their fathers told them to stay the hell out of the trucking business, like mine did, and damned near every driver's hand was raised again. The room filled with laughter as we realized that not one of us had taken our dad's advice.

I am sure my father offered this advice because he knew how aggravating the trucking profession could be. He understood the nature of trucking, that just when you think things are going great, unseen forces always throw the proverbial "wrench"—whether they are flat tires, lights going out, hoses bursting, bad weather, or those cursed weigh stations that all truckers hate. Most truckers have lived at the mercy of these tough breaks and know damned well that these events are going to continue dogging them. Evidently, we are all gluttons for punishment.

So why do we do it? Non-trucking folks are always asking why we drive trucks if we complain about it so much, and it's a fair question, but I say, let 'em scratch their heads and wonder why. You can't understand trucking until you do it—the views, the lifestyle, the rush. Vacationers and businesspeople see some of the great US and Canadian landscapes while traveling, but only truck drivers get to enjoy the grandeur from high up in their cabs. While crossing bridges, the tall concrete walls and Jersey barriers prevent four-wheelers from having marvelous views of the lakes, rivers, or gorges they're crossing. Truckers can watch the shifting landscape from their thrones.

Try to imagine the view a truck driver gets while driving across Staten Island at daybreak as he crests a rise in the highway. The sun, in all its enormity and fire, perched dead center between the two supports of the Verrazano-Narrows Bridge. I've seen views like you wouldn't believe while topping the hill on I-70 West in Hancock, Maryland, about a mile before the intersection of I-68. Just after midnight, halogen highway lights glitter off the bare limbs of apple trees. It's poetry, really. The whole orchard covered in sleet and freezing rain. An ice forest, etched forever in the mind. The road bears a certain beauty, sometimes most evident in the quiet hours and remote stretches that truckers are privy to every ride. Long hauls might inhere long nights and early mornings, but they also inhere access to a seldom-witnessed world.

Truck drivers also have bragging rights from having learned to persevere through rides that would paralyze other drivers. Imagine coming down Jellico Mountain, north of Knoxville, Tennessee, in a freezing fog so thick you can't see anything ahead but a very faint ticker of white lines on the road. You can't see what's behind you, and you have your four-ways flashing to warn drivers approaching the rear of your truck. You can't even pull over on the shoulder—you can't even see the shoulder—but even if you could, you fear another truck will think you're still traveling and hit you from behind. What thoughts race through your mind when you finally emerge from the fog at the bottom of the mountain, when you turn and see the four-inch-long horizontal icicles sticking straight back from your side-view mirrors? You wipe the sweat from your brow. You might even have to change into a new pair of pants. Maybe you add one last verse to the litany of prayers you offered the whole way down the mountain. (Truckers probably pray more in their cabs than in church.) But you survived, and you will next time too.

One of the first things a new truck driver learns, the lesson that's most important, is how to navigate around some of the, shall we say, *less experienced*, four-wheel drivers we all know and love. Most drivers of four-wheel vehicles don't think they're doing anything wrong when they pull in front of a big truck just moments before traffic comes to a screeching halt. Perhaps they're unaware that they did something wrong, perhaps they're inconsiderate or blinded by road rage, but the action is careless and dangerous, and we see more of this behavior every day. Many four-wheel drivers seem to not notice trucks, acting as if four-wheelers are the only ones on the highway. I don't think they realize when they piss off a truck driver, and I think they'd be aghast to know that truckers have several near misses because of their shoddy driving. Take my word for it when I say that it's the four-wheelers causing mayhem on the road. Truck drivers are paid professionals, while many car drivers still need a hell of a lot more practice. Until all four-wheel drivers become proficient at driving, which we doubt will ever happen, truck drivers will always be the more responsible ones—ever mindful of that carload of kids who have the misfortune of having their mother behind the wheel.

Once, while traveling one of Ohio's secondary roads to pick up a load, I arrived at an accident scene just after a trucker saved the lives of a carload of kids. Ohio state police had stopped traffic at the intersection, and curiosity drove me to get out of my truck to inquire what happened. The mother had run a red light, and rather than wiping out the small vehicle, the trucker had somehow kept control of his flatbed load of steel and steered to the right of the car. When I got there, the rig—what you may call a tractor trailer, a semitrailer, or a semitruck—looked like the driver had driven straight into a muddy cornfield for about one hundred yards. Many instances like this cause truck drivers to lose control of their vehicles and the rig jackknifes, with its trailer facing one way and its tractor another, like a pocket knife. The tires were almost completely buried, and the bottom of the flatbed was sitting on top the mud. As I drove past the scene, the woman and kids were laughing, crying, and hugging the truck driver who'd saved their lives. There is no telling how many wreckers—or tow trucks—it took to pull that rig out of the mud.

Truckers develop thicker skin every time they experience a narrow escape, including the *damned near* incidents they are thankful to have survived. I know I've had my share of them. But we do our best, and while I realize most people don't know a thing about trucks, other than that they haul your goods around and, if—or when—you notice them, likely scare the hell out of you because they're so big, you should know that no truck driver heads out on the road thinking, *Let's go out and terrorize the four-wheelers today!*—even if it does have a nice ring to it. No, each truck driver hits the road with the same goals as you: to reach their destination safely and then return home to their family. I will say it again—truck drivers are the professionals of the highways. While you drive to work, a trucker's drive *is* his work and we're not slacking off behind the wheel, we're putting the time in and getting the job done.

And we have to. America depends on truckers for nearly everything. We rely on them to haul our food to our local grocery stores. Our favorite snacks and beverages, and indeed everything that sustains us, doesn't simply materialize out of thin air on shelves and in freezers. And those packages filled with new clothes, housewares, and books that appear on our doorsteps with such convenience? Truck drivers ensure their swift arrival. The home you just moved into? Truckers hauled the beams and bricks. The brand-new washer dryer? The stainless steel refrigerator? You guessed it. Truckers carry the foundations of our infrastructure, too—hauling the supplies that comprise roads, bridges, hospitals, and more. To borrow a couple of the fine, apropos lines of the American Trucking Associations, "Without trucks, America stops," and "If you got it, a truck brought it."

In the pages that follow I'm going to offer a peek into our world. There might be wonkiness, swearing, and a good bit of grease, but you'll get a behind-the-scenes view of how truckers are the bedrock of America, and do a damned fine job of keeping our country humming right along. By telling these stories, I am neither traveling down new highways, nor am I breaking new ground. Every trucker has also traveled these same roads, and each remembers their own special stories. Our stories touch every emotion from belly laughs to tearjerkers, and I'm sure other truckers think of theirs often, as I do mine. The fondest memories that comprise this book—whether hilarious, heartbreaking, or just plain stupid—are threads that, together, weave a tapestry of the American trucking culture.

PART ONE

AN EDUCATION

My two brothers, my sister, and I grew up just a few hundred yards away from our grandparents' farm, and we spent most of our free time there. Our mother had us two years apart, Betty Sue being the oldest, then me, Earl, and Yates. Betty Sue enjoyed cooking and canning with our grandmother, but we brothers had to be outside. Our granddad Obie was a self-made jack-of-all-trades. He purchased tractors and trailers, performed maintenance and repairs, and had a furniture delivery business with my dad. He also watched over a 168-acre farm, where he raised dairy cows, horses, chickens, pigs, dogs, and cats. For a while he also had a huge bull, Jonathan, whom he was very fond of, or at least he was until the day Jonathan chased Obie and my dad up a tree. Jonathan was gone just a few days after that.

Obie owned all kinds of farm equipment, and by the time my brothers and I were ten or eleven years old, we began driving and operating farm tractors, farm trucks, backhoes, bulldozers, and tractor trailers. One of Obie's farm trucks was a red Dodge with two enormous headlights the size of medium pumpkins, one on each fender. One brother would drive the truck and the other two of us would straddle the fenders and ride them through the fields as though we were taming wild horses, and the men would load bales of hay onto the back of the truck. My favorite of Obie's vehicles was a bright red Farmall M farm tractor. After kicking it out of fifth gear, you could fly downhill so fast you could barely hold on, two small steering tires violently shimmying.

No kid should have been doing what we did. Hell, no adult should have. It's baffling that we reached adulthood with our eyes intact—or, for that matter, with anything intact. But the times were great. I'll always fondly remember the monster mud fights that started after a field had been plowed, or a few times I *accidentally* took pot shots at my brothers with a BB gun. And there were great haystack jumps, epic ones really, even if there was that time a hidden pitchfork pierced my calf and got stuck.

The farm was also where I learned to speak like a trucker. I was only five or six when I picked up the first words from Obie. I was watching him handle the Farmall M farm tractor at the time. It was equipped with an electric starter, but the battery was always dead so you had to turn the crankshaft with a handle inserted into the front of the tractor to get it going. The crankshaft is a steel shaft that runs the length of an engine and drives an engine's pistons up and down. The safest way to hold the handle was to push it down with your openhanded palm, rather than grip your palm around the circumference of the handle. You had to hold it just right to avoid a powerful kickback that could result in broken fingers and arms, and Obie was trying to do this, but it just wasn't working. After repeated efforts proved unsuccessful, he stepped back, hitched up his britches, and yelled, "Goddamn you son of a bitch! I wish you would blow up and go to torment!" I'd heard swear words before, but this was real cussing, a whole torrent of expletives, anger, and release, and it was great.

When I got home that afternoon, I stood in front of my little pedal tractor, pretended I was trying to crank it, and yelled, "Goddamn you son of a bitch! I wish you would blow up and go to torment!" It took all of ten seconds for my mother to teach my backside that I should never use those words again. It was all right for Obie to use all the bad words he wanted, but I was never allowed to repeat any of them. Believe me when I tell you that watching my mouth was a very hard thing to do, since all the men in my life—my dad, all the men who worked for my dad, Obie, and all the men who worked for Obie—hell, every grown man I knew—punctuated every thought or sentence with at least a couple cuss words.

I practiced swearing to myself so much that I could cuss with the best of them before I made it out of elementary school. For the most part I kept it to myself, but just a couple times I must have let go, because I remember going to the principal in the fifth grade because Mrs. Gray thought I had uttered an inappropriate word. It must have been some version of "damn," since I remember writing it a hundred times on her damned chalkboard.

I picked up my knack for storytelling from the truckers that Obie and my dad worked with through their furniture hauling business. At any one time, Obie owned between five and ten single-axle tractors and ten to fifteen trailers, so there were always plenty of guys around. Monday through Friday each week, my dad picked up shipments of new furniture at all the manufacturing plants in western North

Carolina. He unloaded the shipments into his warehouse, and then on Saturdays, and usually half-days on Sundays, the shipments would be loaded for delivery to Illinois, Indiana, and Michigan, with each trailer containing twenty to forty stops worth of goods. The drivers who delivered these multi-stop loads were referred to as "stick haulers," and the wild stories they told my brothers and me added a great deal to our "liberal education."

When the stick haulers returned to the farm at the end of a run, or at the end the week, it was time to perform maintenance on the equipment. Obie always told us that equipment wouldn't last long if it wasn't properly greased, so each weekend we brothers attacked every grease fitting on the tractors and trailers. Just like I will never forget my first whiff of diesel, I will never forget the sweet odor of that grease. It was different than anything I'd ever known, a musky scent of dark and thick molasses. Funny how those smells stick with you.

We never had air grease guns, which use compressed air to disperse the lubricant. Instead we had to continually fill hand pump grease guns that were much harder to use, hand-packing Shell multipurpose grease into them from five gallon buckets. Obie made it look easy as he used his index finger to smoothly remove excess grease from the fill end of the gun while rotating it, but it was much more difficult for us. We also had to check the fluid level of each vehicle's rear-end differential housing, the "pumpkin." A drive shaft transfers power from an engine into gears within the differential, which turn the rear wheels on the vehicle. To check the fluid, we would remove the plug and then insert a finger into the hole. If your finger was not covered in oil when you removed it, you had to add more. Unlike the sweet smell of the grease, the differential oil inside the pumpkin smelled terrible, like rotten eggs, and we'd have to violently scrub our hands to get the smell off. We also had the honor of changing each tractor's oil every twenty thousand miles. This required using can openers to open forty-eight cans of Shell Rotella motor oil per oil change, per engine. It was *always* messy, and we would be up to our arms in dirty oil while we removed the drain plug.

Maintenance of each tractor was complete only when its fifth wheel received the proper amount of grease. A tractor's fifth wheel is really not a wheel at all. It is the apparatus attached at the rear tractor frame that allows the tractor to connect to a trailer. To hook to a trailer, a tractor backs under a trailer and the trailer's king pin slides into the groove in the fifth wheel. The king pin causes the fifth wheel's locking mechanism to engage, which keeps the trailer hooked to the tractor. Fifth wheels need lubrication so trailers can turn smoothly. If they don't have enough grease, you need to apply more. If there's too much grease, scrape some off and put it on the fifth wheel of another tractor. If you put too much grease on a fifth wheel, it would get scraped off when the tractor hooked to a trailer, and this excess grease would hang around and eventually bite you. It either fell onto the truck frame, which meant you got it all over your clothes when you greased the truck the following week, or it would fall on the ground and you'd step in it. It also happened that you hardly ever realized you had stepped in it until you either climbed up into the tractor and saw it cover the accelerator, brake, and clutch pedals, or you carried it into the house and ruined the carpet. We had to learn to do things Obie's way, which included a strict fealty to "just the right amount."

Back then we didn't know much about the effect of petrochemicals on the environment. Obie did, however, have a system for recycling used motor oil. After each oil change, we would pour the used oil into fifty-five gallon drums. When a drum was full, Obie would attach his homemade drip pipe to the drum, affix this to the back of the Farmall M farm tractor, and then spread the oil on the dusty gravel and dirt roads of his 168 acres. This kept the dust down, and when traffic traveled those roads or paths, the roads would get packed down and often ended up looking like they were paved with asphalt.

As part of our work, we would hook just-serviced tractors to trailers and take them to the shop, where we would check lights, tires, mud flaps, brakes, and roofs (just to make sure a driver hadn't "forgotten" to mention hitting something and tearing a hole in the roof). The repair of a burned-out light could be as simple as replacing a bulb or as involved as replacing an entire length of corroded wiring, from the front of the trailer all the way to the rear. If there were rusted screws, it could prevent a marker light from maintaining a proper ground, so you would have to scrape rust away from the frame and replace the screws with new ones. Marker lights are small lights on the front, rear, and sometimes sides of vehicles, and they usually work when headlights are turned on.

It's a wonder my brothers and I didn't get killed changing tires. We didn't have air impact wrenches, which use air pressure to loosen the lug nuts, which hold the lug fasteners that secure the tires, so we had to begin the process by breaking the lug nuts using a lug wrench and a three-foot-long bar. Next, we would jack up the wheel by placing a heavy piece of wood under the jack, so the jack would not sink into the dirt or gravel and there'd be some stability. The long bar provided the extra leverage needed to loosen the very tight lug nuts.

After removing the lugs and lug nuts, we would place the three-foot-long bar under the tire, and pull it in such a way that it raised the tire an inch off the ground, allowing us to pull the outer tire off the drum and roll it out of the way. After removing the spacer, a four-inch wide steel ring used to separate the tires, the process was repeated with the inner tire.

Using a valve change tool, we would remove the tire valves—which let air in and out—to let air out of the inner tube; unfortunately for us, tubeless tires had not yet been invented. Once the tire was completely deflated, the real fun began. The tire rims, on which tires were mounted, were comprised of three pieces and referred to as split ring rims. The largest was similar to the tire rim on a car, and there were two narrow rings that fit over the end of the rim, which essentially made it one rim. A combination of differently shaped tools was needed to break loose the rings to separate the rims from the tires, and often included a sledgehammer. Separating the rims from the tires sometimes required a hell of a lot of effort due to rust buildup. The rims were made of steel and if one had not been broken apart for a long time, it had likely rusted so badly that it had become more like one piece than three pieces.

After removing the two narrow rings, you would then lift one side of the tire to remove it from the larger rim. One of the tire tools had a flat one-and-a-half-inch wide blade on one end. After working the tool's flat end under the inner ring, you could hold the ring up until you placed another tire tool next to it. Then, using both tools in somewhat of a leapfrog fashion, you could eventually get the inner ring to release from the outer ring. There were times that required placing the flat end under the inner ring, prying up the ring, and then beating the hell out of the flat end as you worked it around the ring.

With the inner ring removed, you would lift off the outer ring, and then lift one side of the tire to remove it from the rim. Once again, if rust had formed on the rim next to the tire, things were more difficult and a bit of dexterity was required. Standing the tire up on its tread, you would have to hold the tire with one hand while you swung a sledgehammer with the other. Of course, your goal was to hit the rim hard enough to break the rust away from the tire. It usually took just a few well-placed blows to separate the rim from the tire. When you finally got the tire off the rim, you would remove the hard rubber flap, which provided a protective barrier between the tube and the rim, and then you could pull the deflated tire tube from inside the tire. My brothers and I had quite a few laughs watching one another swing the sledgehammers. When you missed the rim and accidentally hit the side of the tire, the hammer would bounce off the rubber and back at such a speed that it would fly out of your hand.

Regardless of the reason for changing a tire, whether it was flat or worn out, it had to be reassembled using a tube and a three-piece ring. To do this, you would place a new tube, or a patched one if the tube was in good enough shape to be reusable, inside the tire and, in the process, make sure there were no foreign objects in there. If there was a flat tire in which we were unable to locate a nail or screw, we would rub its inside with a piece of cheesecloth, which would snag and identify the culprit. If the size of the hole in the tire tube was so small that we couldn't find it, we would overinflate the tire tube and then use a soapy rag to coat the tube. The leak would cause the soapy water to bubble, and then it could be identified and patched.

After placing either a repaired or new tube into the tire, and then the flap, the tire would be placed over the rim. The outer ring would then be set on the rim, and the smaller ring placed inside the first ring. Using a sledgehammer, the rings were beaten into position (or submission). Sometimes, however, one or both of the rings would just refuse to seat properly. This would become very clear during inflation, after a new tire valve was inserted and tightened, when—and usually without any warning that it was about to happen—both rings would explode off the wheel at an alarmingly high speed.

The split ring rims were also known as "widow makers," or "suicide split rims," and older drivers might remember how dangerous they could be. Once, when I was traveling cross-country in a Chevrolet Camaro with a friend, we stopped for gas at a truck stop in Utah. Since this was back in 1970, medieval times, attendants pumped gas for patrons, and while sitting by the pumps, I looked in the rearview mirror and saw a young man inflating a truck tire with a three-piece rim. The guy was sitting on the edge of the tire while using his hand to hold the air hose onto the valve. As I turned to tell my friend that the fellow should inflate the tire differently, we heard a loud bang. I got out of the car and saw that the rim had broken apart and the young man's arm was badly broken. I'm sad to say that I also know of a company that had a tire changer killed when a separated rim blew him up into the rafters of the building in which he was working.

Because Obie knew the danger of these rims, he taught us to lay a heavy steel bar on top the air hose to hold it in place, turn on the compressor, and then get the hell out of the way. He eventually welded steel tubing together to make a cradle that held each tire while it was inflating. For obvious reasons, other tire changing facilities also made tire cages part of their standard operating equipment.

After a tire was inflated, and before a valve cap was screwed onto the valve, the final test was to cover the top of the valve with an extremely high-tech liquid: your own spit. If the saliva didn't bubble from the air pressure, then you knew the valve was properly tightened. Looking back, I suppose it might have made better sense to use some of the same soapy water we used on the tire tubes, but the spit method was more fun. We would then roll the first tire to the brake drum, place the long bar under the tire, and raise the bar while, at the same time, pushing the top of the tire onto the brake drum. After sliding the tire all the way back into position, we placed the four-inch spacer on the drum, then used the same method to place the outside tire on the drum. After inserting the lugs into the bolts attached to the drum, we would finish by placing the lug nuts. We'd hand-tighten them to stay in place, and then using the four-pronged lug wrench and the long bar, we would tighten them further, alternating from side-to-side, until all nuts were moderately secure.

Obie insisted that all tires, whether on tractors or trailers, be perfectly balanced. Wobbling caused tires to wear out more quickly, the steering wheel to shimmy, and the cab to shake, and also produced an aggravating road noise, the sound the vehicle makes as it travels. There isn't a single trucker older than fifty-five who hasn't observed another driver trying to hold onto a steering wheel as it shimmied because of unbalanced steering-axle tires. Chances are that most truck drivers have experienced it firsthand. I know I have.

Back then we didn't have the type of lug-less wheels that don't need balancing, like there are today, so we learned Obie's art of tire balancing. As there was a hole in the middle of the four-pronged lug wrench, we would insert the three-foot-long bar through the hole and then with the lug wrench standing up on two legs, insert one end of the bar into the hole. The other end would rest on the ground, which held the wrench in a slanted position (resembling the jacks of the children's game). The tip of the bar would be placed against one side of the tire. As we rotated the tire, the bar would show which section of the tire was out of balance, because there would either be space between the tire and the tip of the bar or the tire would push the tip of the bar out of the way. By tightening the lug nuts at the location where the tire tool was pushed away, we would continue rotating the tire while tightening however many lug nuts were needed to achieve a perfectly balanced tire. Of course, this made us feel accomplished. Plus, we wouldn't get one of Obie's ass chewings.

My brothers and I could expect to receive an ass chewing when we either did something stupid, such as putting too much oil in an engine due to losing count of the number of cans we'd opened, if we goofed off more than we worked, or if we committed the error of forgetfulness. Obie was usually a patient instructor when he taught us the mechanical necessities of keeping tractors and trailers in tip-top shape, or the proper way to hook the milking suction apparatuses to cows, or how to ride a horse, or how to cut, bale, and stack hay, but then, it was up to us, and in our best interest, to remember what we'd been taught. He would ream us out if we forgot the right way of doing a job.

All of Obie's tractors were equipped with a tire jack, a tire tool, a heavy piece of wood, and at least one mounted spare tire, which was carried in the spare tire rack under each trailer. Back then it was common practice for a driver to change his own flat tire. In the event that one of Obie's drivers sustained two flat tires (assuming the incident happened within four hundred miles of home), Obie wouldn't pay what he called "highway robbery" by covering the cost of a driver purchasing two tires while on the road, and would instead load two tires into the trunk of my grandmother's Cadillac, strap down the trunk, and haul ass to Kentucky or Tennessee. She gave Obie hell when he used her brand-new Cadillac Sedan DeVille for loading tires just two days after he gave it to her as a birthday gift.

Worn-out tires usually got hauled off to the county dump, but my brothers and I had our own ways to get rid of them. We would hook three tires together with a chain and use the Farmall M to drag them several hundred yards to the top of a hill where the hayfield met woods. At this juncture, a road cut through the trees and wound down for about three-quarters of a mile, stopping at a forty-acre delta field at the flank of the Catawba River, and we'd roll them down the road.

Heaven only knows how many truck tires are still in those woods, because we had countless contests to see who could roll their tire farthest. The tires gained a lot of speed as they headed down toward the first curve, and due to that curve having enough banking, the majority of the tires made it through the curve without toppling over. After the tires went out of sight, we listened to them make their final revolutions as they tore through, or tore down, the rhododendrons, bushes, and small trees. Some tires rounded several curves and traveled impressive distances. The tires that fell on the first curve—or before reaching it—lived a second life because we retrieved each one and helped it finish its run.

I'm certain we must have sent more than one hundred tires down that winding road. We should have gone into the woods and retrieved them, but we had other contests to win and lose, so we mostly forgot about them. After Obie and my grandmother passed away, the farm was sold, and I have often wondered what the new landowners thought about all those tires in their woods. Did the woods become a garbage dump? Did flooding cause the tires to float around the woods? Whatever thoughts they had, or may still have, I would wager that no one has figured out that all those tires are there because of three brothers and their adolescent games.

In the years before Obie began installing engine block heaters, the bitter cold of winter made it a difficult time to maintain the tractors and trailers. When it was very cold, an engine's motor oil became so thick that it was damned near impossible for the batteries to crank the trucks. Obie, however, was not one to be foiled by the elements, so he employed his homemade engine warmers. To make the warmers, he used an acetylene-cutting torch to cut the middle out of used fifty-five gallon steel oil drums, leaving eight inches on both the top and bottoms of each drum. We would then pour several inches of diesel fuel, kerosene, into each drum, and then place a rag into the fuel, making sure a small portion of each cloth was sitting above the liquid. Too much rag caused a hotter fire and a big flame that would melt some of the rubber hoses. After sliding a drum under a truck's oil pan, we would light the cloth and let the heater go to work.

Obie's method worked extremely well, and it took no more than twenty minutes to heat the oil warm enough for the truck to start. The biggest drawback to using these warmers was the soot that found its way into each truck cab. We quickly learned to take a hot, wet, soapy rag with us when we jumped up into each truck to start it, and

Obie's drivers were very appreciative that we cleaned the steering wheels and seats before they arrived to leave for their trips.

We also had occasions when we had to deal with trucks that wouldn't start because of faulty air starters, which were more popular back in the day and used compressed air instead of an electric motor to start an engine. The air starters worked well when air pressure didn't leak out of the vehicle's reserve tank, but in the sixties and seventies, small-scale trucking hardly ever ran smoothly, and air leaks were a common occurrence. When a truck wouldn't start because compressed air leaked from the tank, we used one of two methods to crank the engine. The first was to pull another tractor close enough to hook both tractors' glad hands (air hoses) together. Akin to sharing the air from your diving tank with someone who has lost their oxygen, this method would provide enough air for the air starter to do its job. If this exchange of air proved unsuccessful, another method was executed that made use of "the hill" on Obie's farm.

The hill was an extension of Obie's driveway: four acres of grass and hard-packed slick red clay. When Obie owned just a few tractors, they were all parked in the lower of his two barns. While the upper barn was used for hay and horses, the lower barn, formerly a working barn used for dairy making, had been converted into a full-fledged repair facility, with a fourteen-foot-high door for tractors and trailers to go through. As the business grew, Obie ran out of space, and the majority of the tractors were parked up on the hill. When hard rain gouged ruts in the clay, we would fill them with cinders from Obie's coal-fired furnace. The cinders didn't last very long, but until rain washed them away, they added traction to the slick red clay.

Every time a tractor was parked on the hill, it was standard practice to place the gearshift in reverse before shutting it down. Leaving it in reverse effectively prevented a tractor from rolling forward. If there was a problem with getting the tractor started, we sprayed a shot of canned ether, starting fluid, directly into the air cleaner, or into the air intake tube, which was usually located on the opposite side from the exhaust muffler on the back of each truck cab. Then, because ether evaporated very quickly, we would jump back into the truck, depress the clutch, release the parking brake, and as quickly as possible, pull the left gear shift out of reverse and work it into a high gear, ideally sixth or eighth. Most of these early tractors had two-stick duplex transmissions, and it was also standard practice to park each tractor with the high/low range stick, the right one, in the high-gear position, which allowed the engine to turn more revolutions than it would have been able to if it was in a lower gear. In this way, we would get the tractor moving down the hill and after gaining some speed, we would let out the clutch, and usually the tractor would start after a few engine revolutions. I say "usually" because if the hill was too muddy, or snow-covered, we would sometimes travel halfway down Obie's driveway before we gained enough traction to get the engine running.

The trick when using the hill method to start a truck was to build up enough air pressure for the air brakes to function. It was crucial that the brakes worked properly because of the fairly busy country road at the end of Obie's driveway. "Scared shitless" is the best way to describe the feeling of coming most of the way down the driveway with no brakes. Fortunately, we had also scared most of the country road drivers shitless over the years, so they knew to be very alert when they saw a tractor bobtailing—driving without a trailer attached, which means it has very little traction—down Obie's driveway, especially during the winter months. This was a small town where everyone knew one another, and most of the time, drivers, both men and women, would congratulate you with a thumbs-up if the truck successfully started.

If the truck did not start on the first attempt, we would give it another good shot of ether, push in the clutch, and roll the tractor out of the driveway onto the paved secondary road, which had enough of a downhill slope to gain momentum and crank the truck.

Modern truck parking and braking systems make it impossible to crank a tractor by rolling it down an incline, but this tactic is unnecessary because present-day engines start easily in even the harshest, coldest weather. We older drivers have graphic memories of being greeted by the sound of our truck engines barely turning over when we needed them to start. If we had to stop, knowing our truck was not likely to start up again because of the weather, we always tried to park on a downslope, but driving through flat country usually made this impossible. If your truck wouldn't start, you had to either ask another driver to pull your rig with a chain or call a mechanic for service. I don't remember too many drivers pitching fits or raising hell when this occurred. It was just how things were back then. Eventually you started your truck and you went on about your business.

Today's trucks are greatly advanced compared with those of days gone by, because they're equipped with air dryers as a component of their air braking systems. Most commercial vehicles are equipped with air brakes, and rubber hoses, known as "airlines," carry compressed air from a tractor's air compressor to all brake assemblies on both the tractor and attached trailer. In freezing weather, it's necessary to release built-up water and condensation, which would otherwise freeze airlines, making brakes become inoperable. If you've ever been at a stoplight beside a tractor trailer, or even a school bus, and heard a loud expulsion of air, then the air dryer was blowing out built-up water and condensation. But before the advent of this technology, truck drivers and mechanics had to open the petcock (drainage valve) of each air tank, which were positioned on the bottom of the tanks, in order to purge the water and oil mixture that accumulated in each tank because of the weather. Some petcocks had wires or cables attached to open them, and this made it much easier to drain the tanks than it was when you had to squat under each tank and twist each petcock open with your hands. You also had to try to keep your feet as far away from the air tanks as you could, because the liquid was so nasty that it ruined most everything it sprayed.

If drivers forgot to drain their tanks several times each day in freezing weather—or if they were simply too lazy to do so—the slimy, gray mixture would work its way throughout both the tractor and the trailer's airlines. Depending on how badly the trailer brakes were frozen because of this, there were various tools employed to thaw them, such as blowtorches, hot water, and kerosene heater fans. If a driver was resourceful, as most were, he would use an emergency flare, three of which are part of each truck's roadside safety equipment. These red flares, the kind you notice at an accident scene, can burn from ten to fifteen minutes and reach temperatures of up to twenty-nine hundred degrees Fahrenheit. You could also get your brakes to release by pouring isopropyl, or denatured alcohol, into the tractor airlines. After hooking the airlines to the trailer, and then charging them with air, the alcohol would work its way back through the trailer airlines, usually reaching and defrosting the brakes.

Of course, some drivers were unable to fix their brakes themselves and had to call a road service company. For them, fixing frozen brakes brought in good money during the winter. They were often summoned to thaw brakes that had frozen while the drivers slept in rest areas and truck stops.

Over the years, Obie employed many truck drivers. During the Bay of Pigs invasion in 1961, I asked him why he was sending the drivers out on their runs on Sunday, when it looked to me like we would all be blown up by an atomic bomb. He told me that he wasn't going to bury his head in the goddamn ground because of what other people were doing. He said he couldn't alter the outcome of the situation, so the trucks were going on their runs as planned. It was a life lesson: Be your own man and make your own decisions.

For good or bad, Obie's drivers, like most truck drivers today, also lived according to their own compass. Many truckers, especially owner-operators, (O/Os)—drivers who are not employees, but contractors responsible for their own road expenses—are independent and free spirited, living like they're the last American cowboys. When you drive a truck, you spend a great deal of time being your own boss. When away for any length of time, whether a few days or a month, drivers make most of their own decisions, and as long as they perform their jobs satisfactorily, no one gets in their face. But this preference for individuality also has a way of making drivers into floaters. Driver turnover rates were high back in the 1950s and 1960s, and they still are today. I remember Obie running through quite a few truck drivers. Most of them quit after a few months if Obie hadn't already fired them. I wasn't old enough to think much about the fact that they moved on so quickly, or why, I just enjoyed meeting all the drivers who came and went. Many of them made lasting impressions on me, and I learned a hell of a lot about trucking from these fellows.

Some of Obie's drivers left keys to their personal vehicles while they were on a trip, in the event their car needed to be moved. When my brothers, Earl and Yates, were teenagers, they would keep the batteries charged in some of the drivers' cars by driving them all over the county. Several of the cars were what would now be called "muscle cars," or powerful, high-performance vehicles, and, evidently, my brothers flexed them pretty hard. I was either in college or Vietnam when they were driving these cars, but hearing them tell their escapades made me long to join them. I remember Earl telling me that he drove most of the drivers' cars, including a 1964 Ford Galaxie 500 Fastback 390 V8 and that same driver's 1956 Chevrolet pickup truck, which had a very large engine. He said it would burn the back tires in first *and* second gear. He also drove a 1956 station wagon full of tools that would fly all over the place as he sped around curves.

It seems that some of the drivers told my brothers that they knew their cars were being driven, and not just around the block, because when they returned the owners would find that their cars had considerably less gas than they did when they'd parked them. Well, some of these truckers must have been just as adventurous when they were kids, because they told my brothers to just take care of the vehicles with a knowing wink.

One of Obie's truck drivers was a man I'll call Clete. He was a tall, lanky, handsome fellow who always smiled, and word had it that all the truck stop waitresses loved him. Clete always got to work on time, and he usually arrived at Obie's early enough to wash his tractor before he left on his trip, that is, if we boys had not had time to wash it. (Earl and Yates usually washed Clete's tractor as a thank-you to him for being one of the guys who probably knew they were driving his personal car—in Clete's case, a fast Ford—while he was out of town, but didn't object.)

As a kid, I spent most summer nights at my grandparents' house, and I was awoken one morning around four o'clock by a loud engine roar, indicative of a diesel motor with no muffler, which kept getting louder and louder. I looked out the open window and saw Clete's dark green Mack H67 truck turning into Obie's driveway. Obie was also roused by the noise, and he soon learned that Clete had found a "torch man" at some truck stop, a guy who specialized in giving vehicles this type of sound effect. The fellow had cut the top off the truck's muffler, removed its baffles, the sound muffling chambers inside a muffler, and then braised the top back into place. The gutted muffler on the Mack was very, very loud, and those were the days when having the loudest truck really meant something. The loudest truck turned everyone's heads, and a driver was awfully proud when his truck turned heads at truck stops.

Clete grinned like a Cheshire cat when Obie agreed the truck "sounded like a truck ought to sound," but the grin faded fast when Obie told him the muffler would have to be replaced. Obie considered it too unprofessional. To our amazement—especially Clete's—Obie did not take the new muffler cost out of Clete's pay, but he did warn Clete not to gut any more mufflers.

Another reason I remember Clete was because he once returned from a trip with the front of his trailer roof peeled back several feet. His story, which he stuck to, was that a highway had been repaved and the extra pavement raised the trailer so high that it bounced up and the corner of it hit a bridge. Clete was not the only driver who destroyed the top of his trailer, especially since we had many weekly deliveries in downtown Chicago, but none of the other drivers ever thought up an excuse better than his. The overwhelming majority of trailer roofs were damaged when drivers tried to drive under low overpasses. (The City of Chicago had many low overpasses, mostly due to its elevated train system.) When I received my chauffeur's license at eighteen, the damaged roofs provided me with ample driving time because the trailer repair shop was about eighty miles away.

Another driver, Harley, was a beanpole-skinny country fellow with a high-pitched voice. He was a decent driver during warm weather, but feared driving in the snow. If he knew it was going to snow during a trip, he would call in sick rather than take a chance getting stuck driving in it. On a trip from western North Carolina to Illinois, he hit snow flurries when he reached Cincinnati. He spent the night at a Cincinnati truck stop, hoping to find clear roads the next morning, but instead found several inches of snow, so he hopped on a Greyhound bus and went home to North Carolina. Obie had to send another driver to get the rig and make the delivery in Illinois. Later, when Harley begged Obie to give him another chance, Obie told him to come in the next Sunday and he could go back to work. On Sunday, Harley came to work, did his pre-trip inspection of his tractor and trailer, and was ready to leave the yard when it began blowing light, flakes of snow. He grabbed his belongings, got out of the truck, and told Obie he couldn't head out because it was snowing. I don't remember all the cuss words Obie flung at Harley, but he certainly told Harley he was fired.

When spring arrived, Harley called Obie and asked if he was looking for any good drivers.

Obie's replied, "Yes, I am, but I don't know where in hell I can find any."

We never saw Harley again.

Another one of Obie's drivers that I remember was Tom, a local, middle-aged man, who had worked quite a few jobs over the years, so I suppose his stint as one of Obie's drivers just added to his reputation as a Renaissance man.

I played Little League baseball with Tom's son, Tommy, who was one of our small town's celebrities. One winter, after an extremely heavy snowfall, many of the local kids were sledding down a long steep hill on a busy stretch of highway. I'm not sure what was on the rails of Tommy's sled that day, but the sled moved like greased lightning and he went flying past the spot where everyone else's sleds had stopped and picked up even more speed going down another small hill. He was going so fast that he was unable to stop for a red light. His sled zoomed underneath the belly of a tractor trailer that drove through the intersection at the same time. We all cheered, whooped, and hollered when he emerged from under the trailer unscathed; we'd been sure he was going to get squished under its tires. The truck driver remained none the wiser throughout the whole ordeal, and Tommy certainly beat the devil that snowy day.

Tommy's dad, Tom, didn't have his son's good fortune, and one time returned from a trip in his Emeryville International truck with the entire right front of the tractor demolished. Tom said that a deer had been loping across an open field beside the highway and just as he was driving by, the deer jumped the fence and landed on the highway in front of his truck, which was going sixty miles per hour. Some of the metal parts of the truck (back in the day, there was more metal than plastic) had remnants of deer hide and hair embedded in them, so it was a believable story. This wasn't the first time, and wouldn't be the last time, a tractor came home with deer damage, and times haven't changed much in that regard. Deer strikes are still damaging scores of trucks each year. I think it's fair to say that truckers are doing their part to help reduce the deer population.

The fact that Tom had many types of employment had a lot to do with his frequent alcohol consumption. When I was seventeen, Tom asked if I wanted to ride with him to pick up a load of furniture in Asheville, North Carolina, about thirty-five miles away from Obie's farm. I had nothing better do to, so I hopped into the truck's passenger side and we headed out. As soon as we were out of sight of Obie's house, Tom pulled over at a wide spot and asked if I wanted to drive. I had occasionally driven a truck over the past several years, but I didn't yet have my chauffeur's license. I happily accepted anyway, crawled behind the wheel, and began driving up the mountain to Asheville. As I tested all ten of the truck's gears, Tom relaxed into the passenger seat and opened a pint of Hiram Walker's Ten High bourbon. While we inched along in the traffic of Asheville's Tunnel Road, he flirted with every woman we passed. (I learned an awful lot about flirting that day.) By the time I'd picked up the load, driven us back down the mountain, and dropped the trailer at our warehouse, Tom was shit-faced.

It wasn't long before Obie had to let Tom go. Obie knew that Tom was drinking and driving, and that he wouldn't put the bottle down anytime soon. If he kept at it, it was only a matter of time until someone got hurt.

And Tom certainly wasn't the only driver who had a strong relationship with the bottle. My father was a man who enjoyed his bourbon, and I learned early on how to detect a man's alcohol level by his breath, and there were many times I knew a fellow had been drinking before he uttered his first word.

Far be it for me to know why Obie tolerated this as much as he did. I can only assume it was because he was always in need of drivers. There was a dearth in the industry then, and there still is today, which means that a good interview with the owner has always been likely to guarantee a man a trucking job. There were no drug or alcohol tests required at the time (today they're routine), and a man's ability to hold off from the bottle for at least a short time was measured by showing up sober for his initial interview. Past employer verification hardly ever happened.

After a driver began working, his job performance could quickly betray his blood alcohol level. If he didn't show up for work and didn't call to let Obie know he wouldn't be in, odds are he was drunk. If he showed up for work and was staggering from his personal vehicle to his truck, he was drunk. If he returned from a trip with a damaged tractor or trailer and no good excuse for the damage, yep, you guessed it: probably drunk.

Other than truck drivers, Obie usually only had one man who worked around the farm. We called him "Bus," although I've never known whether that was his given name or a nickname. He was working there when I was born, and stayed on the farm for many years. I remember him still working there when I left for college. He was a man of small stature—probably didn't weigh much over 120 pounds, and stood around five feet three inches tall—and was similar to most respectful men of the time, removing his hat in the company of a woman and holding the door open for others.

Obie gave Bus a few dollars every week, but only paid him all his wages once every couple of months, and Bus never seemed to have a problem with that. When money *was* in his pocket, Bus would always stop by the local bootlegger's house before he went home, and get good and liquored up. The alcohol must have made Bus think he was a big man by the time he got home, because once there he would commence beating on his well-over three hundred-pound wife, Sarah Belle. The sheriff's office was usually summoned when this happened, and Bus would be sent to jail. He would work as part of a chain gang for thirty days, and on day thirty-one, Bus would come walking up Obie's driveway, with his hat in his hand. He would find Obie on the back porch and Obie would just grin at Bus and tell him to go on and get his ass back to work.

One spring morning, I stood on the drawbar of the farm tractor, the beam across its rear end, and held on as Bus drove it down to the river bottom to disk the forty-acre field in preparation for corn planting. As we rode, I smelled last night's liquor on Bus's breath, so it was no surprise when, as we reached the field, he said he wasn't feeling too good and was going to go over *yonder* and rest just a minute before he began. I nodded to indicate I'd be fine by myself and watched as he headed into the woods to find a cool spot. His "minute" of recuperating turned into an afternoon. We usually went back to Obie's for lunch, but that day I picked some carrots, green onions, and cabbage from the vegetable garden beside the field and ate them instead.

Later, just as I finished disking the field, Bus returned. He must have been watching me because he came out of the woods just as I was wrapping up, and he was itching to go home. He crawled up onto the drawbar, where'd I been that morning, and I drove us back to Obie's house.

When I was fifteen, Bus gave me some advice that I didn't fully understand until years later. It happened one afternoon while we were stacking hay in the upper barn. We were talking about women, and my girlfriend, and Bus began pontificating about men and women having sex. He stated that when a man is finished with the session, most women are ready to go again, and that they are hard to satisfy. In a way that only Bus could explain this situation, he said, "Mista Ed, when it comes to women, I just want you to memba one thing . . . a woman can always look up longer than a man can look down!"

Sarah Belle was a wonderful woman who often worked for my mama. She did laundry and ironed and babysat my siblings and me when we were very young. One time, when my brothers and I were all under ten, we rode with our dad for about five miles from his furniture warehouse to our home. We were in an H model Mack truck equipped with an air starter and as we bobtailed home along the two-lane country road, the tractor's air starter produced loud, high-pitched sounds that could be heard for miles. While slowly rounding a curve, our father moved over a bit because Sarah Belle was walking along the side of the road. As she waved to us, our dad reached under his seat and pulled up the handle on the air starter. This sudden, ungodly noise scared poor Sarah Belle so badly that we watched in the side mirrors as she jumped over a ditch and climbed her way up a small hill beside the road. She later told us that after the initial surprise, she laughed about the episode so hard that she cried.

Shortly after I secured my chauffeur's license, Obie said he wanted me to go somewhere with him and instructed me to go up on the hill and pick any tractor I wanted to drive. I chose a single-axle H model Mack with a ten-speed, two-stick duplex transmission, did a pre-trip inspection to confirm all looked good, and then we bobtailed several miles away to see one of Obie's friends who owned a construction company. We borrowed a two-axle lowboy trailer from Obie's friend (a lowboy is designed to carry heavy equipment and vehicles, and ours was equipped with eight or ten chains and binders), and then headed west on I-40.

We drove through Asheville and then southwest toward the Nantahala National Forest. The forest is in western North Carolina, and the road through it is about as mountainous as it gets. We traversed a treacherously curvy US Highway 74 for many miles, and then Obie instructed me to turn onto a wider-than-usual gravel road. The road wound through the woods for close to a mile, and then opened to reveal an immense rock quarry. I soon learned the purpose of our trip was to pick up a Caterpillar D6 bulldozer that Obie had purchased from his friend.

Picking up a bulldozer is no simple feat. Moving heavy equipment on the highway typically requires a permit, sometimes escorts, and, most importantly, the tractor and trailer carrying it must be able to legally transport the weight of the equipment. Because we were driving a single-axle tractor and a two-axle lowboy, it was legal for us to carry a little more than a thirty thousand-pound payload, but the dozer was more than thirty-six thousand pounds—too heavy. We decided to go ahead anyway, and drove it onto the lowboy. We should have removed the bulldozer's ten-foot-wide blade, and come back for it later, so we wouldn't be over the eight-foot width limit, but we decided to leave it attached (although we did angle the blade as much as we could). Since the weight was illegal anyway, I guess Obie figured, *What the hell, let's go for it.* Almost the only thing we did correctly was to use every available chain and binder to secure the dozer.

All Mack trucks have a steel, miniature bulldog on the front hood, and as the tractor strained as we went up and down those mountain roads, I kept imagining Mack's bulldog turning around and asking, *Why are you treating me this way?* But it was Obie's show, so I told the bulldog to quit bitching and do his job. (I had heard my dad comment—many times—that you know you are pushing a Mack truck too hard when the bulldog turns around to look at you with tears in his eyes, so maybe I got this image from him, though he usually had too much speed, not too much weight.)

I felt certain that we were in serious trouble when we reached the very open North Carolina weigh station on US 70 at Swannanoa, mostly due to us being overweight, too wide, not having enough axles, and having no permits. As I pulled into the facility, Obie told me to park to the side. He hopped out of the truck, hitched up his britches, and strode into the office like he owned the place. Twenty minutes later, he walked over, climbed into the truck, and said, "Don't go over the scale. Just stay to the right of it and let's go home!" I never learned what this "bypass" cost Obie, but I have always wished I'd been a fly on the scale-house wall while Obie was stating his case. It turned out that all my worrying was for nothing.

I was a happy camper for the next ten miles, but then we reached the top of Black Mountain on I-40. The road there is treacherous and unfriendly for six miles, terrorizing all truck drivers who happen upon it. Even though I kept the truck in a very low gear, we hadn't traveled more than a half-mile down the mountain before the brakes on both tractor and trailer were smoking. Obie told me to keep my foot on the brake pedal, continue to apply pressure, and not take it off, as this would have allowed air to reach the very hot brake shoes and possibly cause them to catch fire. The farther down Black Mountain we went, the more smoke poured out from the brakes. And the smoke screen was so dense that the no driver could see well enough to pass us—they had no idea what lay ahead. After what seemed like hours, we finally made it past that last curve at the bottom of the mountain, and Obie instructed me to take the Old Fort exit. I'm sure the good folks of Old Fort remembered us for quite some time after that visit, because the smoke-shrouded truck left the smell of burned brake shoes in its wake as we traveled down Main Street.

When we finally arrived in Obie's driveway, I shut off the truck, and then Obie looked over at me and nodded his head. It was as much of a pat on the back as I would ever get from him, and it was all I needed.

Forty-plus years ago, there were no signs giving instructions on the mountain, trucks did not have to pull over at the top of it, and there were no truck escape ramps. Today, North Carolina law requires all eastbound trucks to pull over and stop at the top of the mountain to inspect their vehicle's brakes for proper functioning; the speed limit for trucks is thirty-five miles per hour; four sand-filled truck escape ramps punctuate the road at one-mile intervals; and most of the time, at least one of North Carolina's finest parks somewhere down the mountain to monitor—and strictly enforce—the speed limit. The truck escape ramps were built in 1979, just a few years after Obie and I were there, following an incident in which a tractor trailer's brakes stopped working when it was coming down the mountain. Its driver managed to steer it to the final curve, but then it shot through the median in excess of one hundred miles per hour and wiped out a carload of Western Carolina University kids.

One afternoon soon after the safety features were in place, I drove by after a rig had gone into one of the sand-filled ramps. It did such a good job that the tractor had come to an extremely abrupt halt, but, unfortunately, the trailer's momentum kept it going fast enough to tear the fifth wheel off the tractor's frame and slam into the cab. The trailer broke the locking mechanism on the back of the cab and the two hinges that connected to the frame on the front of the truck. When I got there, the driver was standing beside the cab of the International CO-4070, which was cleanly sitting upright on the sand in front of the trailer, as though a forklift had placed it there. I think the Department of Transportation experimented with several different sand and gravel mixes, but they finally got it right.

Under Obie's tutelage, my siblings and I learned much of what we needed to help us lead happy, fulfilling lives. He instilled a sense of pride in us when we did something correctly, and taught us how to make things work, to fix what was broken, and even rebuild when necessary. Earl certainly used this life skill when he drove his VW Beetle home from the Marine Corps base in Parris Island, South Carolina, and wound up with an expired motor. Fortuitously, it died just as he was passing a gas station, and he was able to roll it into the parking lot. He asked the owner if there was an auto parts store in the vicinity, and, quite unexpectedly, the owner gave my brother directions to the store and tossed him the keys to his pickup truck. Utilizing tools supplied by the gas station owner, Earl dropped the motor from the frame, rebuilt it, reinstalled it, and drove on that same motor back to western North Carolina.

Learning to be a trucker incorporated a lot of trial and error, and more often than not, our trials ended in error. One morning, Obie told me to bobtail the five miles to our furniture warehouse to retrieve a certain trailer that needed repairs. Rather than take the shorter route along a curvy two-lane country road to town, I took another route, so I could access the three-lane US 70. I can't say for certain why it was that I wanted to take US 70, but it was likely so that I could drive a little faster than I would have been able to on the curvy road. When I reached a stop sign, I pulled up behind a fellow in a pickup truck who was also going to drive east on US 70. When traffic cleared, we both began to merge onto the highway, traveling about forty-five miles per hour. After we had driven close to one mile, I glanced at my side mirrors and then looked back straight ahead, only to see the pickup rapidly slowing in front of me. I applied the brakes—hell, I locked them up—but my bobtailing tractor didn't even slow down. (The antilock braking system was still on the drawing board.)

While sliding, I turned the wheel to the right and damned nearly missed the pickup, but the left front of my tractor caught the right rear of the truck. The rear of my tractor spun around and destroyed the mailbox of a nearby muffler shop, and the impact shot the pickup to the left with such force that it flew across two oncoming lanes of traffic. Somehow, by the grace of God, he missed every car. After a massive sigh of relief, I jumped out of the tractor and ran across the road to check on the driver. He had gotten out of his truck and was inspecting the damage. He said the truck belonged to his brother and that he would go inside the service station to call him to come have a look at the damage. I asked him if he was physically okay, and he said that the impact had not been too hard and that he would be just fine.

I then crossed back over the highway to where my tractor was sitting sideways in front of the muffler shop. The shop's owner, whom I knew, told me he put the mailbox in the back of his shop after I hit it. He said no one needed to know that I had destroyed federal property.

A North Carolina highway patrolman soon arrived, gathered the facts, and then issued me a citation for following too close to the pickup. He seemed to understand my side, but said he was bound by law to give me a ticket. The pickup driver's brother arrived while the patrolman was writing my ticket, and I watched as the brothers carried on a very serious and very animated conversation. They were still discussing the issue when I followed the patrolman across the road to them. Speaking on his brother's behalf, the pickup truck's owner said that the driver had neck and back injuries. The driver, who had been fine until his brother's arrival, suddenly suffered from great pain. An ambulance showed up to haul him to the hospital for what turned into a two-week stay. I wasn't surprised, though, when a nurse friend of mine who worked at the hospital informed me that there was nothing physically wrong with him, but he'd insisted on staying anyway. Knowing most everyone in our small town had its advantages.

Obie surprised the hell out of me by saying, "Oh hell. These things happen." When my court date came along several months later, Obie surprised me again when he gave me a blank check to use to pay whatever fine I would incur.

When I took the stand, I went through the events up to the moment I turned to observe the pickup stopping in front of me. I also testified that I did not observe any brake lights or signal lights on the pickup. The patrolman corroborated the fact that neither sets of lights on the pickup were in working order. It was decided that I was not at fault, and acquitted. (The judge and his wife frequently played bridge with my parents, but who's to say if that made a difference.) When I handed Obie back his check, all he could do was grin and say, "I'll be damned."

Months later we learned that Obie's insurance company had settled with the pickup truck driver for $5,000. This was for the driver's alleged pain and suffering. His brother's pickup must not have experienced as much pain and suffering because the owner never fixed the minimal damage, and we often saw it around our small town. From then on, I always drove slower and was more alert when I was driving trucks, especially while bobtailing, but I was only eighteen at the time, and there were things about trucking and life that I had to learn the hard way.

PART TWO

MILITARY TRUCKING

In 1966, conventional wisdom dictated that I study hard and stay in college, but those were the days of the Vietnam War, student protests, and the free-spirited culture that would soon lead to Woodstock—an opportune time, I thought, to release myself from the antiquated shackles of attendance and hit the beach, which was a short ninety-mile ride from the school. In July 1967, the college suggested, via letter, that I take the next semester off to reassess my desire to do well when I returned to classes.

During my imposed hiatus I returned home and while at the barbershop one morning, I noticed a poster depicting military construction workers on bulldozers and other heavy equipment. The men resembled John Wayne or Dwayne "the Rock" Johnson: strong, courageous, pinnacles of masculinity. It was a Navy poster, an ad for young men to join the Seabees. I presented myself to the Naval recruiter the next morning to inquire about joining.

He asked when I would be drafted, and I told him I wasn't sure, but that it would probably be in the next few months. He said that I should go find out, so I walked down the hallway and entered the local Selective Service office. Mrs. McIntyre, a friend of my mother's, clerked the front desk, and I asked if she knew when I would be drafted. She took a moment to collect herself, and then said, "Honey, do you see that pile of letters on that desk? Well, they are going out tomorrow, and I am truly sorry, but your name is on one of them."

I politely thanked Mrs. McIntyre, walked out into the hallway, marched back down the hall, entered the recruiter's office, and asked where to sign. He told me that they would put me in the regular Navy if they couldn't successfully billet me as a Seabee and I told him hell no. I'd rather become an Army ground-pounder, a member of the infantry, than be stuck in an ocean a thousand miles from land. He assured me that he would take care of my request, and I soon enlisted in the United States Naval Construction Battalions, the Navy Seabees, the unit of the military comprised of builders, mechanics, electricians, engineers, equipment operators, steelworkers, and utilities workers.

Because I grew up around trucks, farm equipment, and other heavy equipment, I was able to enter the Direct Procurement Petty Officer (DPPO) program, which recruits workers with existing technical skills and starts them at a pay grade that is higher than other entry level positions. I was assigned the title of Equipment Operator (EO), indicating my occupational specialty, and my rate on entrance was as a Petty Officer Third Class, which has an E-4 pay grade. In the Navy, enlisted personnel have rates, not ranks. A person's rate is determined by their pay grade and reflected in the number of chevrons, the V-shape markings, on the person's rating badge, which is worn on the sleeve of their uniform. The rating badge also shows their occupational specialty, indicated by the symbol above the chevrons. Those who didn't enter the Seabees under the DPPO program referred to us as IPOs, or Instant Petty Officers.

I am certain that having a trucking background saved my life when I served in the military. If I hadn't become a Seabee, my Vietnam experience would have most likely entailed wading through leech-infested water and worse. But I had the good fortune to join the Seabees, and I thank God every day that my name did not become one of the 58,318 inscribed on the Vietnam Veterans Memorial wall.

Several days after I signed on the Navy's dotted line, I was sent to Fort Jackson, an Army base close to Columbia, South Carolina, where I was inducted into the armed forces. I can't say that I enjoyed parading around naked with hundreds of other guys while we suffered the ignominious examinations of every orifice during our physicals. One doctor asked if I had any physical limitations, and I told him that two years earlier my right eye sustained a broken floor of the orbit and five broken bones had to be removed and replaced with a plastic plate that now supported the floor of my eye. He asked if I could see okay or if I had blurred vision. I said I was fine, so he declared me "fit as a fiddle" and I was shuffled off to the next specialty doctor. It was very late in the day when they completed poking, prodding, listening, and embarrassing us, and sent us to eat what can only be described as tasteless vittles with a bunch of Army recruits.

I lay awake that evening questioning my reluctance to have read more while in school. I wondered if all the fun I had in high school was worthwhile or if I should have studied beyond the bare minimum. I questioned the year in college that I had wasted at the beach and many other choices I made prior to my induction at Fort Jackson. But feeling sorry for myself wasn't going to do a bit of good and I would never avoid the draft by fleeing to Canada, so I thought, *Just suck it up. You made this mess of shit, so lie in it!* The thought gave me some ease and I soon fell asleep.

When I received my written orders the morning after induction, I learned that I was going to Gulfport, Mississippi, for four weeks of boot camp. I felt a jolt of excitement—I assumed I was going to fly there from South Carolina, and I had never been on an airplane. But I was wrong. I was destined for the road again. This time packed onto a Greyhound bus for a twenty-hour slog south to Mississippi.

On arrival, we suffered through buzz cuts and were issued uniforms, underwear, and many other items, which, when properly packed, completely filled up the single duffle bags we had been issued. Since we wouldn't need the civilian clothes we had worn to boot camp, they gave us boxes and Uncle Sam paid for our civvies to be sent back home. When we attended our first classroom instruction the following morning, the base's commanding officer welcomed us and then proceeded to inform us that he knew we had all joined the Seabees to stay out of Vietnam—and to be fair, Seabee literature showed that they were stationed all over the globe, including places like Alaska and Italy, so I'd hoped to end up somewhere like that. He said that while he applauded our fine decisions, he could guarantee that every goddamn one of us was going to Southeast Asia. The man must have known what he was talking about because within six months, we all shipped out.

The four weeks of boot camp flew by. We learned to properly make our bunks—yes, to have a coin bounce when thrown on our bed—how to fold our clothing, and how to pack the other items we'd received. We had to practice packing until we knew how to fit every article inside our duffle bags. We spent most of our days in various classes, learning about the Seabees. We were not going to go out and kill people, but we needed to learn how to stay safe amid the violence, including how to manage the possible assault of tear gas. For practice, we had to don gas masks, and then enter a fifteen-square-foot building filled with clouds of tear gas. Our instructions were to take a very deep breath with the mask on, then pull it off and count to ten before running outside; they wanted us without the mask for ten seconds so we could feel what it was like to be in a tear gas attack. Those ten seconds were awful, and most of us could not even count to one before we choked and scrambled out of that hellhole. On the first try, the instructor must have not spoken clearly, because many of us pulled our masks off and *then* took a deep breath. Those who screwed the pooch had to do a practice maneuver a second time, so, by God, we learned to take the deepest breath before removing our masks.

We were inducted into the Seabees to build things, but not because we were all in tip-top physical shape. To shape up, we marched some on the grinder, which was a large, paved asphalt lot, maybe five acres, and had the occasional run. When they made us run, quite a few of us dropped like flies. The Marines and Army drill sergeants would have punished those who dropped out, but most of our instructors were older inductees and couldn't have cared less. Another part of our exercise required that every now and then, probably twenty of us had to lift a telephone pole onto our shoulders and walk a couple laps around the grinder. Twenty men and one pole? Light as a feather.

When boot camp ended, I finally got my airplane ride, taking an Air Force Lockheed C-130 Hercules transport plane from

Gulfport to Ventura, California, which was the closest airport to our Seabee base in Oxnard, California. During military training in California, I drove one of the troop-hauling tractor trailers, or cattle cars, as we called them, when we went into the hills each day to play soldier. The hills of southern California were quite small compared to those of Black Mountain, so I wasn't at all bothered by the drive and thankfully never experienced the "pucker factor," which was probably best described by a guy named Frank I met in later years. In his words, the pucker factor is when you get so scared that "the ass puckers so tight that you couldn't drive a nail up it, even with a big hammer!"

After our training in the hills, when the troops would board the cattle car to travel back to our base, some of them would ask me to try to make them fall over, so I would gently rock the steering wheel from side to side as I drove. We may have been training for men's work, but many of us were still kids at heart.

The average age in our battalion was between eighteen and twenty-one, but there were at least a dozen older fellows, guys who had enlisted because they needed a big change in their lives, such as getting away from a nasty divorce, or because of a court order to either join the military or spend a couple years in jail. Most of the latter were hard drinkers who never smiled or smiled too often. (We tried to stay away from the creeps.) The older guys tended to be a pain in the ass to us younger soldiers, but the ones who were also rated EO sure as hell knew how to operate bulldozers, graders, backhoes, front-end loaders, and scrapers. Some of the older fellows who were good at their jobs became good instructors, mentors even, imparting their brass tacks wisdom to many of us younger equipment operators.

One part of our military training was learning how to throw hand grenades. On our second day of this training, a fellow soldier didn't throw his grenade far enough over the wall and a small sliver of steel flew back at us and got stuck in my arm. My inquiry asking if I was eligible to receive a Purple Heart fell on deaf ears.

We also learned the proper use of a Colt 45 handgun and M16 assault rifle. I had learned to shoot shotguns and rifles while growing up, but the M16 was unlike anything I had fired before. During my first training to use the weapon, I learned to hold the gun down, so the muzzle wouldn't walk toward the sky. The Marine gunnery sergeant was quite adamant about us holding it down since there's no point in shooting above your target.

All of us had to qualify in the proficient use of each weapon, and my file noted that I qualified well, good enough to become a sharpshooter. I'm sure I could have gotten an even better mark than this, an expert level mark, as the day before qualification day I shot at the expert level and did so on many other days too, but we were permitted to visit the enlisted men's club the night before qualification day, and that California beer must have messed with my eyesight. I'm not clear what happened to the guys, and there were many, who were unable to shoot straight enough to qualify. Probably nothing, since we all went to Vietnam anyway.

We spent several months getting familiar with new models of earth-moving equipment, and attending more classroom instruction, the majority of which taught us how to interact with, and be tolerant of, the citizens of South Vietnam. We were warned what not to do, what not to do, and what not to do. We were not to disrespect the Vietnamese people. We were not to go into any unauthorized area not protected by the US military. And we were absolutely not to have sex with Vietnamese women.

After we completed training, they let us travel home for a few days, and on our return, injected us with enough vaccines to protect us from every disease known to man. We felt like pin cushions, although we realized the necessity. Then, just a few days later, we boarded airplanes and headed west to the war. The experience started with a bang. At Vietnam's Da Nang airport, we were walking down the plane ramp and onto foreign ground at the very same time that two fighter jets lifted off the tarmac and unleashed a roar that scared the shit out of me. I instinctively hunkered down, sure the place was getting hit with rockets, mortars, or something else that would take our lives before we even got going. No one called me a sissy for this, and no one pointed at me and laughed. Many of us reacted this way.

Without any real threat, we entered the Da Nang airport terminal to retrieve our duffle bags. We were dressed in clean utility uniforms, and our boots were all spit polished. Everything about us was green—our uniforms, our experience, our complexions. We'd been drinking heavily for most of our twenty-four-hour plane ride. I don't remember who happened to be the highest-ranking officer on our flight, but somehow, he knew most of us had smuggled bottles of fine bourbon, gin, vodka, and scotch onto the plane. He had to punish us for breaking the rules and sneaking in contraband, so he forced us to forgo the mitigating graces of proper cocktail mixers. I can thank the awful combination of fine bourbon and orange juice for my sorry state.

Walking through the huge Da Nang terminal, I saw a dozen Army soldiers sitting on the floor and leaning against a wall. The fellows looked the exact opposite of us Seabees. While we were fresh and wide-eyed, with clean, barely used uniforms, they were dirty. They wore no socks in their jungle boots and clearly hadn't shaved for a long time. They didn't look directly at us, or at anyone else. All of them had what we soon learned to be the thousand-yard stare. They didn't smile or frown, and didn't speak to one another or us. They were silent and displayed no emotions whatsoever. I didn't think I had lived a sheltered life up till then, but this sight was very unnerving and unlike anything I'd ever seen. I'm sure they were headed home—and I badly wished I could have gone with them.

Our tasks in Vietnam would include rebuilding Highway 1, building numerous bridges and buildings, rebuilding airstrips, and cementing landing pads for helicopters, among other endeavors. For my first six months there, I was one of two Seabees who had the best job in our battalion of 850 men. For at least four days each week, I drove tractor trailers hauling supplies from the central, coastal city of Da Nang, which was Vietnam's biggest city and had the second largest military installation, to our base camp, Camp Haines. It was a journey of only seventy-five miles north of Da Nang, but it was slow. Prior to our battalion, Mobile Construction Battalion 10 (MCB-10), renovating Highway 1 into a forty-foot-wide super highway, the road was a muddy path about one-and-a-half lanes wide, which meant that you had to damned nearly stop each time you encountered another vehicle coming toward you from the opposite direction. And you had to drive carefully as the road wound through the middle of numerous towns and hamlets, including those of the large city of Huế.

Despite these annoyances, what made it so that we had the best job were the absolutely spectacular views that part of the trip offered.

The road north of Da Nang began at sea level and wound for thirty miles over the North Trường Sơn mountain range. At the top of the mountain was the Hải Vân Pass, which literally means "Ocean Cloud Pass" and is South Vietnam's most dangerous mountain pass. It is some fifteen hundred feet above sea level and the view from there was breathtaking. Waves crashed into the sheer cliffs along part of the mountain, and rolled up onto white sandy beaches toward the bottoms of both sides of the mountain. At fifteen hundred feet, you could also see Vietnamese sampans—flat-bottomed boats used for fishing—and US naval vessels sharing the same waters. I never got tired of making the trip.

If you drove empty, going southbound from our camp, or fully loaded, going northbound from Da Nang, you would travel across a bridge going over a calm, clear blue bay filled with sampans. There was a fifteen-mile ascent to the top of the pass, and the sampans diminished in size each time you saw them as you drove through the numerous switchback curves. The switchbacks also let you see the vehicles that were ahead of yours, so you could be prepared. The higher up you drove, the more you could see the widening of the South China Sea below, and the military naval vessels anchored many miles out, punctuating the water.

At the top of the Ocean Cloud you felt like you were literally on top of the world. Those last few miles before the summit, you could let yourself imagine, just for a moment, that you had escaped the ugliness of the war. But you were slapped back to reality at the top of the mountain when the sight of the concrete and sandbag bunkers dashed your quiet serenity and brought you back to reality. On the drive down the other side of the mountain, if you kept your eyes focused left, toward the ocean, the drive was similar to the ascent, but when you looked straight ahead, your view was overtaken by Da Nang.

Guardrails had been installed on the majority of the road's ocean side, but it didn't take long for me to see that they weren't very effective. Most every trip revealed new evidence of a vehicle having recently crashed through the guardrails. Due to the sheer drop-offs, practically every vehicle would have hurtled toward the ocean at the bottom of the mountain. I couldn't help but imagine, depending on where the vehicles broke through the guardrails, the drivers' horrifying final rides. I just hoped it wasn't other servicemen who were plunging to their deaths.

On our initial trips south from our camp to Da Nang, the other driver, Baker, and I were surprised to find all traffic stopped at the bottom of the north side of the mountain. We learned that this was because it was deemed too unsafe for single vehicles to travel across the mountain—the risk of enemy fire was too great—so traffic would be held to allow an entire line of vehicles to move as a convoy. Armored personnel carriers and jeeps equipped with machine guns were interspersed throughout the convoy to provide some semblance of security, and every day, one convoy traveled north and one traveled south, typically in the midafternoon. If drivers arrived early, they would have to wait in line until it was time for the convoy to proceed.

One of the biggest drawbacks to having to travel up the mountain as part of the convoy, at least in my case, was that because it took an entire day to make the seventy-five-mile trip, by the time we reached Da Nang, we had always missed the northbound convoy. This meant we had to wait until the next day to head back to Camp Haines; down one day and up the next day. The silver lining was that this allowed us to spend two nights each week in Da Nang, which had abundant shopping, numerous forms of entertainment, and excellent Army, Navy, and Marine enlisted men's clubs. This was a far cry from what we had back at our camp in the boonies. The enlisted men's clubs were large and usually had live music. If I was lucky enough to spend a weekend in Da Nang, which could happen if I arrived late on a Friday and my return load wasn't available for loading until Monday morning, I could usually enjoy a USO tour show, live entertainment sponsored by the United Service Organizations. I once missed a Bob Hope show by one day. Most of the USO shows were fantastic, with good-looking, scantily clad girls dancing while musicians performed popular songs. It was all good fun, but it made us think about what we had left behind in our hometowns.

While this time spent waiting at the convoy point was good drunk fun, it didn't allow a quick turnaround, and since Baker and I were the only drivers assigned to the Da Nang supply runs, it was imperative for us to turn around as quickly as possible each time. It would have been better for everyone if we didn't have to wait, so one day Baker and I went to our higher-ups and explained that we lost hours waiting, which they already knew, and pointed out that it had been many months since there had been enemy activity while traveling through the Hải Vân Pass. Our good officers explained the issue to our battalion's captain, and within days, we each had a signed order that allowed us to travel anywhere in I Corps at any time. I Corps was the northernmost part of the four geographic military sections, or corps, that South Vietnam was divided into. It stretched from the Demilitarized Zone (DMZ) north of Quảng Trị to just south of Da Nang.

Shortly after receiving our new orders, Baker and I both happened to make our initial "travel anywhere" journey the same day. At the convoy point, we presented our orders to Army Military Police (MP) soldiers, but they said they didn't care what the orders read. One said, "You crazy-assed Seabees WILL be convoying with them, so get your asses back in line!"

Well, this MP was unaware that a person should never say no to Seabees, or really any Marines. Rather than getting our asses back in line, we requested an audience with that MP's sergeant. After one short radio communication, the fellow was told that our orders actually did mean "anytime, anywhere."

He scowled at us and with obvious disgust said, "Fuck it, go ahead," and then told the other MP to let us go over the mountain by ourselves.

This situation, being out of the ordinary, garnered a lot of attention from other convoy drivers, both military and civilian. As we drove ahead, many of them were standing outside their vehicles, and a number of the guys shouted support for us Seabees, such as "Way to go," "Good luck," or "You lucky SOBs." However, interspersed with the well-wishers were drivers who hurled invectives such as "Hope you get shot, you sorry bastards," "The VC [Viet Cong] will skin you alive," or, our favorite, "Seabees are a bunch of crazy fuckers." I don't know how Baker handled the accolades, but I was smiling and waving to everyone as we drove past them. And we didn't have a single problem as we made our thirty-mile drive up and over the mountain.

With an early start, our new anywhere, anytime order made it possible to make one round trip to and from Da Nang and another one-way trip back to Da Nang before the road closed at dusk. I'm not sure whether we had anything to do with it, but about two weeks after going over the mountain alone, both north and south convoy points were done away with, and traffic went up and down the mountain at any time from dusk to dawn.

There were two drawbacks to not having to convoy across the mountain. The first was that we didn't have several hours to sit, drink beer, and shoot the breeze with fellow soldiers. The second was that it was much harder to relieve yourself while driving because your vehicle was going faster, so you really had to remember to "go" before you headed up the mountain. It wasn't at all difficult to go when your vehicle was going slow, which I assume is due to the Department of the Navy having had the driver's comfort in mind when it selected our tractors.

The tractors were efficient, referred to as "multi-fuelers" because their engines could be configured to burn different types of fuel, and equipped with notched, pull-out throttles, which when set, would allow the truck to maintain its speed without using the accelerator. It was an early form of cruise control that came in mighty handy while you were traveling up the mountain at five miles per hour after having enjoyed numerous beers waiting in the convoy line. As the throttle held the truck at a steady speed, the large, wide running boards made it easy for a soldier to relieve himself when the urge came along. It really wasn't all that hard to drive with your right hand as you stood on the running board and held the door open with your knee. With your left hand free, you could unzip your britches and complete your maneuver. On some of the switchback curves quite a few other drivers could be observed utilizing this system.

You might witness this careful operation less frequently these days, due to the presence of more women in military units. In 1968 and 1969, the only females we saw were a few nurses, girls in USO shows, and occasionally Vietnamese women. We never got called out for using the throttles to aid in this endeavor. It was simply one of the things people were more lax about in Vietnam than they would be in the US.

Vietnam was a lot of things, but dry was not one of them. Cheerful young girls, who were around ten years old, always had an ample supply of beer kept ice-cold in ice-filled holes dug in the sand. I don't remember the price of a beer, but I do remember getting shit-faced on just a couple dollars. I tried to pace myself on that cheap nectar so I would be in an acceptable condition when it came time to drive, but I can't say the same for the other drivers. I witnessed many of them drinking while they were running up and down Highway 1. It was quite possible that some of them were already half in the bag when they arrived at the convoy point. And maybe it was simply due to the thousands of Army trucks on the roads, which caused significant overcrowding, but I saw many trucks overturned or run-off into rice paddies, and I'd venture to bet that drinking often had something to do with that.

Army tank retrievers looked like tanks with attached wrecker booms—tow trucks with the kind of heavy-duty chains and hooks used to secure a vehicle. We didn't have to drive far before we saw the retrievers in action because they were always busy righting overturned trucks. These tank retrievers also came to the Seabees's rescue when one of our fully loaded twin-engine Euclid TS-24 scrapers, a tractor used for digging, moving, and leveling ground, got mired in a rice paddy. Some people said the TS-24 operator got pissed off about something and drove into the rice paddy on purpose. However it happened, the scraper was so badly stuck in the mud that the retrieval required the tank retriever plus our other two fully loaded TS-24s. It was quite a sight and quite a sound: the combined forty-five-hundred horsepower engines screaming and struggling to pull that behemoth from the mud.

I got to know a number of different kinds of trailers in Vietnam that I hadn't had any experience with at home. Except for driving Obie's lowboy trailer, the one we used to transport the D6 Caterpillar bulldozer down Black Mountain, all my early trucking experiences were spent driving forty- or forty-five-foot closed vans. There were no closed vans in our battalion's equipment inventory, so I was relegated to pulling flatbeds, lowboys, or stretch (extendable) trailers. Several flatbeds were equipped with side racks and tarps to be used when something needed protection from the rain or needed to be kept out of sight.

My first flatbed life lesson was provided by an Army colonel during my first trip to Da Nang. My tractor was hooked to a flatbed and I was waiting in line with several other trucks to load some product for my trip back north to Camp Haines when a petty officer told me to move to a different section of the storage yard. He said they had a small job for me that wouldn't take much time. I drove over, and was instructed to haul four pieces of flat steel plates to a fabrication building near the Da Nang base. As it was very close and I would be driving very slowly, I didn't chain the plates down on the trailer bed. It didn't seem necessary. But just when I was in sight of my destination, I turned right at an intersection and the steel plates decided to go a different direction. They shot right off my trailer and landed in the roadway, just barely missing a jeep carrying the colonel and his driver. It was a close call, and *way* too close for comfort. One look from the colonel and I felt like Beetle Bailey as he stood ready to catch hell from Sarge.

I climbed out of my tractor, came to attention, saluted the colonel, and began making my apologies. The colonel must not have had important business elsewhere, because he stayed on the scene while his driver took me to find a forklift so I could load the steel plates back onto my trailer. When I returned, the colonel taught me *his* proper way of securing steel plates for transport. His way was to use all the chains and binders the flatbed was equipped with, and there were many of them. I hadn't used any of them, and even though two chains would have sufficiently held the number of plates I was carrying, he made me put on every single chain and binder. When I was finished, he said, "It's a damned miracle you didn't kill us, and maybe next time you will think twice about not chaining your load." It was an awfully good lesson. I'm pretty sure my life would have turned out differently had I flattened the colonel and his driver.

I also had to quickly learn the rules of hauling a motorized sheepsfoot soil compactor. A few months into our time in Vietnam, we'd heard a rumor that a brand-new one had been issued to our battalion. Both Baker and I wanted to be the one to haul it north because it was rumored to be one aggravating son of a bitch to transport and each of us enjoyed a challenge. Well, I was the lucky trucker who happened to be in Da Nang on the day the equipment arrived. The thing looked like an enormous wingless insect on steroids. It wasn't heavy, maybe fifteen to twenty tons before adding water to the reservoirs in each wheel, but it was twelve feet wide and close to twelve feet high. Imagine an asphalt roller, triple its size, and add hundreds of steel appendages, resembling actual sheep's feet, welded to each roller. The compactor had articulated steering, which meant it had front and rear halves connected by a pivot point, so it was fairly easy to navigate, and it could do some mighty fine dirt packing.

The sheepsfoot had a very small footprint because only a few of the feet of each roller touched the wooden floor of the lowboy trailer, and I didn't have a hell of a lot of confidence the thing would stay on the trailer for seventy-five miles. Heeding the aforementioned colonel's advice, I used every chain and binder I could find on that sucker. I had so many chains holding it down that I don't think it would have moved if the trailer had flipped upside down. It was such an ungodly looking piece of equipment that the people who viewed it, military or civilian, stopped what they were doing and watched it go by. When I arrived at our camp in it, most everyone came out of their offices, or "hooches" as we called them, to see it. Even the mechanics stopped what they were doing and stood in the shop's doorways staring at the spectacle.

One of the equipment operators stationed at our Da Nang base had driven the compactor onto the trailer when I picked it up, so not having a clue how it operated, I stopped at the transportation hooch to ask who would unload the sheepsfoot. The first-class petty officer looked at me and said, "You brought the goddamn thing up here, so you get to unload it!"

It didn't take but a few minutes to figure it out, and after backing down the loading ramp, I got the thing in high gear and headed to the shop for them to give it a physical. I went flying past the transportation hooch and the dust cloud I left pissed off the petty officer, which of course was my intention. What was he going to do? Slap my hand and send me off to Vietnam?

Several weeks later, I was alone while heading northbound from Da Nang with another load of supplies, and as I started down the other side of the mountain, I detected the smell of hot truck brakes somewhere ahead. It was a smell that had been seared into my nostrils and memory when I hauled the D6 down Black Mountain. The farther down the mountain I went, the worse the odor became. After another curve or two, I could see smoke billowing from the red-hot brakes of the responsible tractor and trailer. It was Baker's rig. He typically tried to haul more freight than everyone else who drove tractor trailers, and his heavy load of steel beams had put his brakes to the test.

When I got close enough to see through the smoke screen, I saw Baker steering the tractor with his right hand while his left arm held the driver's side door open. His left foot and most of his body was on the running board, and his right foot was pushing the brake pedal. We were damned nearly at the bottom of the mountain, so I dropped back a short distance to not choke on the smoke. When we hit flat ground, he was able to stop the vehicle on a large, open piece of land, and I came to a stop too. We both got out and I walked to him, but there was so much smoke still billowing out that we had to walk upwind, and quite a way so we could breathe. I asked, "What the hell were you doing coming down the mountain standing on the running board?"

"I kinda figured my brakes would completely go out," he said, "so I pulled the trailer hand valve all the way down [they used to stay in the down position], then opened the door and stood on the running board prepared to jump off. If the son of a bitch was going over the cliff, I sure as hell wasn't going with it."

Baker had grown up in Montana and some of his family members had been loggers. He told me that a couple of them had done this same thing when their brakes were also put to the test, and although the practice wasn't exactly safe, the alternative—running off a mountain—made it the best option. This made damned good sense to me, but I'm glad I never had to employ his method.

Another crucial lesson I learned early on regarded the hauling of one of our most prized commodities. Each week, I made two or three round trips from our camp to the supply depots in Da Nang to haul food, grease, lubricants, and flat and structural steel (well-chained), culverts, and cement. By far the most important load I hauled, which was about once every two weeks, was a load of shrink-wrapped, palletized, double-stacked beer. It wasn't really necessary to cover the pallets since the beer was already skunked from sitting outside in the sun at the beer yard in Da Nang, but during my first beer load heading north, I realized that tarps were absolutely necessary.

On this run, I was winding through a hamlet when I noticed the driver in an approaching Army truck blinking his lights and pointing above my truck. I didn't see anything unusual as I looked into the sky, but my rearview side mirrors revealed several little Vietnamese boys on top the trailer. They were grabbing cans of beer and throwing them as fast as they could to their friends running behind the trailer. I stopped quickly, prepared to get out and holler at them, and the boys skedaddled. When I climbed on top the beer load to assess the damage, I saw that half a dozen cases were missing from one of the pallets. The boys had only thrown one beer at a time, but they'd worked quickly. Practice must have helped them become such proficient beer thieves.

There was nothing I could do, so I got back in the trailer and got going, but within minutes, I saw beer flying off the trailer again. I don't know where the boys came from or how they climbed back on the trailer without me seeing them. This time I wasn't able to pull over quickly, so I grabbed my Colt 45 from its holster, pointed it toward an unattended rice paddy, and fired several rounds. That did the trick. The kids must have thought I was shooting at them because they hightailed it out of sight. For all future beer hauls, I set up the side boards and tarped the beer to hide it, thereby removing the temptation for these little bandits. Interestingly, the ice-cold beer in the sand at the convoy points were the same brands of beer I hauled. I'm sure this was not a coincidence.

For one of my trips in a stretch trailer, I was hauling 110-foot-long telephone poles, which, like the structural steel I hauled on a stretch trailer, were used as bridge pilings. On this occasion, I loaded ten of them at our barge facility in Huế and set out for a trip to the bridge construction project thirty-five miles north. These loads were over twice the length of a flatbed, or lowboy, so it was best to drive slow and steady with them. In a small hamlet I very slowly rounded a gentle curve and swung wide enough for the trailer to clear a villager's hut. As the trailer straightened out, I glanced out the passenger-side mirror and saw an old mangy dog whom I'd observed painfully hob-bling through this village many times before. As my eyes returned to the road ahead, the dog dashed from behind a bush and ran right in front of the trailer wheels. I didn't even have time to react. I stopped as quickly as I could, got out of the tractor, and ran back to the scene, where there was already a crush of people gawking at the flattened canine.

Suddenly realizing that joining the crowd maybe hadn't been the smartest idea, especially since I had just run over someone's pet, I started walking back to my tractor. But then it struck me that the crowd wasn't upset that the dog had been run over; instead, they were arguing over who was going to take the poor guy home for supper. No one seemed to even notice me, so I jumped back into the truck and slowly got the hell out of Viet-Dodge. I felt terrible the rest of the trip to the bridge site. There was no way I could have avoided it, but it still bothered me that I had killed the old guy. I could only hope that someone took him home and had a good meal.

After the dog incident, I got back on the road and continued on until I reached the bridge staging area. I arrived with just enough time to remove the chains and release the standards that bound my load, drop off the delivery, and make it back to our camp's gate before sunset—though this required my driving like a bat out of hell and I just barely made it.

The next morning, the pile-driving crew arrived at the staging area, only to find that none of the 110-foot-long poles I'd dropped off the day before were there. I was sent back to the site and was inter-rogated about them. What happened, they wanted to know, to those motherfucking telephone poles? I calmly explained the work I'd done the previous evening, and even showed them where the poles had gouged holes in the dirt when they fell from the trailer. There were no marks indicating the poles had been dragged away, and there was no sawdust, which would have meant someone had cut them into smaller pieces, so, like magic, the poles had simply disappeared.

Several days later, the mystery was solved by all who drove Highway 1 north of our camp. The poles had been stolen, and the crime committed by the entire population of the closest village to the jobsite. The villagers had assembled enough bodies to lift the long, heavy poles onto their shoulders, and, using what I call the "Vietnamese shuffle," carried them away. (I use the term "shuffle" to describe the way I observed the manner in which most Vietnamese walked. They didn't walk loudly and take long strides like Americans, but quietly shuffled their feet, seemingly taking smaller steps.)

For the next several weeks, we watched the poles be transformed into planks for building huts, as men sat astride them and sawed back and forth with their crosscut saws. I can only imagine how excruciatingly painful their legs and asses must have been after sitting on poles covered in creosote, a tar-derived substance known to swell, irritate, and burn the skin. Since it took weeks to saw these poles, their long periods of exposure most likely caused their skin to develop sores and become sensitive to sunlight. The other villagers must also have suffered problems after breathing the nasty creosote odor, because damned near every hut sported newly sawed creosote siding.

After that incident, we hauled the poles from Huế to a fenced-in area at our camp's storage yard. When the poles were needed at a jobsite, each truck driver would load his trailer using a large Pettibone forklift. The one thousand pound, 110-foot-long poles were tapered from approximately thirty-six inches in diameter on one end to about eighteen inches on the other end, and the big end weighed considerably more than the smaller end. This meant that when picking up a pole, each forklift operator had to place the forks closer to the heavy end. The forks were about eight feet long and when spread to maximum width were probably eight or ten feet apart. Because of the height and weight of each pole, it was impossible to load more than one at a time, so it took quite a while to pull and load your trailer. Being young'uns, we had contests to see who could drive his forklift and place his forks under a pole and balance it on the first try, and also which of the two drivers—there were usually two drivers hauling poles to a jobsite—could load the maximum ten poles on their trailer the quickest. We would do anything, really, to break the monotony.

Just as we learned not to leave poles lying around, we also learned to bring all our equipment back to camp each evening. Every good truck driver knows that it's always best to make as few trips as possible—unless he's paid according to the number of loads he hauls—so each afternoon, we loaded as much equipment as possible onto our lowboys. We weren't too concerned by height, weight, and width—if we could chain it to the trailer, we hauled it. You would see loads that would never be legal in the US. It wasn't unusual for us to load two D6 dozers on the same trailer, or two mini-dozers and one backhoe, or one rubber-tired crane and a backhoe. It didn't matter if equipment stuck out over the top of the tractor, as the boom of a rubber-tired crane did, nor did it matter if equipment protruded far past the end of a trailer, as a backhoe's bucket or rubber-tired crane did. We certainly improved on the meaning of "full utilization," and, surprisingly, none of us were ever called down for overloading our trailers.

All Seabee truck drivers and equipment operators carried holstered Colt 45 handguns and M16 assault rifles. Thankfully, none of us ever fired them during combat, even though we worked in hostile territory almost every day. We didn't have to fire our weapons because the times that we were required to perform construction projects in areas of ongoing enemy activity, or while working at night (to minimize traffic disruption) when replacing highway drainage culverts, we were guarded by either Army or Marine troops. Both groups had our backs, and we will always be indebted to those soldiers.

When an opportunity arose to thank the Army soldiers for providing security, we took it. We'd heard some of them saying they wished they had some of their own dump trucks, like we did, so they could stockpile sand, which would enable them to fill the sandbags needed to build stronger bunkers. Their desire for this was completely understandable since their Camp Evans, which sat adjacent to the Seabees's Camp Haines, was frequently under fire from enemy rockets and mortars. This was likely due to the fact that Camp Evans was an operating base for Cobra attack helicopters, which the enemy wanted to destroy. Understandably, the nightly rocket attacks scared the soldiers. Hell, they scared us too, and the incoming warning sirens also sent us to our bunkers.

With gratitude we spent some of our time off on Sunday afternoons delivering dump truck loads of sand to numerous locations throughout Camp Evans. After we dumped the first load of sand, word quickly got around the base, so all we had to do was drive through their gate and soldiers would appear to direct us to delivery points. One afternoon, we had so many dump trucks on their base that it looked like a Seabee camp.

Some of the Army bunkers were quite elaborate, requiring many sandbags. When they depleted the initial supply of sand, the Army soldiers said they would love to trade us for more truckloads. We had the sand they wanted, and they had things we wanted: jungle boots, boonie hats, and poncho liners, so we were happy to make a trade. We learned the barter system well in Southeast Asia. Not only did we barter with the Army guys, but also with Vietnamese people. In exchange for our C rations, they gave us sunglasses and rubber flip-flops made out of old US tires and tubes.

One night during a mortar attack on Camp Evans, we ran from our beds and dove into our bunker located on the side of our hooch. When we noticed one of our guys missing, a friend and I ran back into our hooch to find him. Bob was trying his best to crawl into his clothes locker. We surmised that his action may have been caused by the effect of his evening at the enlisted men's club, the only bar on our base where enlisted soldiers could drink alcohol. It seemed that Bob had made the most of the establishment and gotten quite inebriated. So inebriated, in fact, that we had to pull him out of his locker and drag him out to the bunker.

The following morning, Bob didn't believe our account of what happened during the night, and although he did remember having been in the bunker the previous evening, he wasn't about to admit that he had tried to paw his way into his own locker.

One of the poncho liners acquired in trading turned out to be an especially well-used possession for a friend of mine. During lunch one day, this fellow pulled his dump truck into position alongside a dirt wall, so he would load first after lunch; at which time, a bulldozer would push the dirt over the top of the twenty-foot-high wall, and a front-end loader would then load each dump truck. He then climbed down from his cab and walked a short distance to greet a nice looking eighteen-year-young Vietnamese woman who had come to visit him, and then welcomed her up into the bed of his dump truck. He climbed up after her, holding several beers and his poncho liner, which had been procured by bartering a dump truck load of sand with an Army soldier, and I assume was used as a blanket to lie on.

I suppose our commanding officer enjoyed keeping track of how we were performing our duties, because he drove up in his jeep shortly after the driver began attending to his lady in the dump truck bed. The commanding officer, the CO, asked the missing driver's whereabouts, and someone told him that he had gone behind some bushes to heed the call of nature. After shooting the breeze with the rest of us for a long few minutes, the CO wished us a good afternoon and drove away.

After lunch, our industrious and elusive driver was grinning from ear to ear. Before we got back to work, the dozer operator hollered down at us from his perch up high to get our attention. He begged us not to call him a Peeping Tom, but confessed to watching the driver with his girl while the CO had been on site. Apparently the dozer's routine was to sit on his equipment while he ate his C rations, enjoying a panoramic view of all the personnel and trucks below him, and on this day he had gotten quite a show. He also said that the soldier who had told the captain about the missing driver doing his business sure as hell knew what he was talking about. The driver had certainly been giving the business to the girl the entire time the CO was talking to us. The dozer operator said he damned near laughed out loud when the dump truck driver looked up at the dozer operator, grinned at him, and shot him a thumbs-up. We all applauded the dump truck driver for *unloading on time*. It turned out that this was not just a one-time fling. That same girl came to meet the lucky driver at one work site and then the next. Love—or lust—seems to conquer all obstacles.

When we were away from our camp and not close to a location serving hot food, our standard culinary fare came from boxes containing various kinds of "Meal, Combat, Individual" rations, more commonly referred to as C rations. Each box typically contained one can of meat and vegetables, such as beans and franks, or spam and a vegetable, but one of them—the absolute worst—was cold ham and eggs, although everyone called them H and MFs (ham and motherfuckers). Another can would usually contain fruit, and other cans, or packages, could offer chocolate, crackers, cocoa powder, cigarettes, or a folding, pocket-size can opener, better known as your basic P-38.

C rations typically tasted pretty awful, although their flavors could be made more palatable if they were heated. Most of the guys who drove trucks, or those who operated heavy machinery, learned that the hot exhaust from a motor's manifold, which funnels the exhaust, functioned like an excellent stove to heat their meals. After puncturing a hole in a can with your P-38, you would then set the can on top of the manifold. Sometimes you might have to leave the motor running so the manifold would stay hot. Other times, it was so hot outside that the engine naturally stayed hot for a long time. Either way, in fifteen to twenty minutes, you were eating a warm meal.

If we weren't away from Camp Haines, we ate at our US Navy-operated mess hall, our "chow hall." Food was served all day from about 4:30 in the morning until 9:00 at night, and there were few restrictions. If you were a soldier, you could eat as much, and as often, as you liked. This was wonderful for us since Navy cooks have always had a reputation for serving the bests meals of any of the armed forces, and our cooks reinforced this reputation. Each morning, I looked forward to breakfast because three or four cooks manned separate stoves and every imaginable breakfast fixing surrounded each stove. You could ask for an omelet with any ingredient, or pancakes, waffles, or eggs fixed any way, or steak and eggs, or almost any other breakfast you could dream up. Our chow hall received an award for being the Best Chow Hall in I Corps. It was a hell of an honor since we were pitted against the much larger and better supplied chow halls in Da Nang.

It was my job, and privilege, to haul at least half our food supplies from Da Nang. I had heard that our mess hall's procurement officer—who was probably a master chef—excelled at the Navy's art of comshaw, a bribe or payoff. It may have been that he was just a damned good barterer, but the foods he secured (by whatever means) allowed us to eat what honest-to-goodness tasted like home-cooked meals. It was light years better to eat in our mess hall than be stuck eating those damned H and MFs.

With so much good food available, and numerous, varied choices, it often happened that we weren't able to eat everything we piled on our plates. But our uneaten or partially eaten meals did not go to waste. After we placed our trays at one of the kitchen windows, the kitchen crew removed the utensils, napkins, and other inedible items, and the leftovers were scraped into washed out fifty-five-gallon steel drums and left outside. Shortly after each breakfast, lunch, and dinner, and on a daily basis, our closest Vietnamese village's food envoy could be seen leaving our camp with one, sometimes two, of the drums mounted onto his cart, which had an axle and two passenger car-sized tires.

On my regular Highway 1 drives, I would watch what took place after the envoy arrived at his hamlet with our leftovers, and I was astounded the first time I saw it. A number of villagers would come out to meet the envoy, and then one of the villagers used a large ladle to scoop out food to fill the pots and pans that residents brought with them. It didn't seem to bother them that different meats, vegetables, breads, and desserts were mixed together. None of us Seabees could have fathomed having to choke down that concoction, and I am sure every one of us would have become violently ill if we ate from these drums. But then, I doubt that any of us have ever been as hungry as they were. The Vietnamese seemed to relish each meal as if it had been freshly prepared, and we never heard of them getting sick from the food.

There was, however, something that regularly made me sick as all hell. It came in the form of the weekly malaria pill all US soldiers were required to take. They were horse size, close to the size of two quarters stacked together, and they must have worked, since I don't know anyone who contracted the disease, but the pill's side effects were not fun: nausea, vomiting, abdominal pain, upset stomach, headache, diarrhea, weakness, and loss of appetite. They didn't make me vomit or lose my appetite, but within hours of swallowing one, the rest of the side effects would start to wage war on one another.

It began with an upset stomach, a bit of nausea, and a slight headache. Soon after came a wave of abdominal pain, the likes of which would damned nearly bend you double. This pain would *always* be immediately followed by violent diarrhea—a side effect that could be tolerated when in close proximity to a privy—but try driving a truck down Highway 1 when the symptoms take hold. To distract yourself, you think nice thoughts of being back home, or try to sing a popular song. Your body straightens out, stiff as a board, and you squeeze your cheeks very tightly. There were several times that I had to bring my rig to a screeching halt in the middle of the highway, jump out of the cab, and suffer the embarrassment of squatting by the side of the road while nature took its course. I can still remember the calls from the Army dudes driving by my vehicle. Many blew their horns at me. Others even hollered, "Been there, done that." Oh yes, all truck drivers knew to keep a roll of toilet paper in their trucks.

The last of the side effects, weakness, caused me to slowly pull myself back up in my truck, where I would sit for several minutes while taking long, deep breaths. It was rough, but I know the pill's side effects were mild compared to the suffering I would have experienced had I contracted malaria. Fifty years later, when the occasional stomach bug comes to visit, or when preparing for those dreaded colonoscopies, I am mentally transported back to the memories of having to take those pills.

By far though, the worst part of my job in Vietnam happened each morning at daylight. I always drove one of the first vehicles on the road following the Army ordinance crews (minesweepers) clearing of Highway 1, which they did before traffic was allowed onto the roadway. On these early morning drives I'd see dead bodies laying in the middle of the road at villages, hamlets, and towns. The South Vietnamese Army soldiers placed the bodies of the Viet Cong, or suspected Viet Cong, and sympathizers killed the previous night in the roads leading into each community. The bodies were supposed to make a statement about what would happen to those who aided the enemy. There might be one body lying alone, or a pile of bodies placed on top of one another.

I remember driving very slowly by the bodies the first time I saw them. I didn't get sick or anything like that, but it took me several days to get the sight out of my mind. The sight of them was certainly a hell of a way to start each day. They were a reminder that I was in the middle of a war zone. I never got used to seeing them, but I saw them so often that I began to accept the practice as just another one of the many unnecessary things that happened during my time in Southeast Asia. War, as they say, is hell.

After driving trucks for six months, I asked to be taken off the Da Nang runs so I could become more proficient at operating heavy equipment—bulldozers, scrapers, and loaders. After I did this for two weeks, I was asked if I would volunteer to do a special job for the Army. One might think that only a fool volunteers for anything while in military service, but I was quick to volunteer; I thought this offered a chance to get out of camp for a few days to do something unique. It wasn't more than a couple of days later when I watched as a Sikorsky S-64 Skycrane helicopter hooked an International mini-bulldozer to it and then took off and flew westward into the A Shau Valley, west of Huế. I then boarded a Huey helicopter and off we went in search of the Skycrane. The Skycrane flew slower than the

Huey and we eventually passed it and landed at an Army firebase on top a mountain.

The sight there was surreal: completely devoid of vegetation. Soldiers had pointed Army artillery downhill and fired shells at all the trees and shrubbery until nothing was left but dirt and rocks for several hundred yards. They'd made it so that the enemy couldn't sneak up on the firebase since the land was barren and there was nowhere to hide.

Soon after I disembarked from the Huey, the Skycrane set the mini-dozer down on the mountain. Over the next two days, I operated the dozer to dig a ten-foot-wide by fifty-foot-long by ten-foot-deep hole on the mountainside. Within an hour of my finishing the hole, another Skycrane flew in to place a pod into the hole. The pod was eight-feet wide by forty-feet long by eight-feet high, and was complete with all the comforts of home. Its purpose was to house some commanding Army colonel or general, although I never found out who stayed there. I do know that after spending two nights in a mountaintop bunker, ensconced in a sleeping bag beside twenty Army soldiers, I was ready to go back to camp. As I grabbed my boots to get ready to head out, one of the Army guys told me to be sure to shake out the scorpions. Thankfully, my boots didn't have any of the critters, but I watched the others pour them out onto the ground and smash them with their boots.

When the Skycrane eventually returned to take the mini-dozer back to Camp Haines, I got to experience the static electricity discharged by hooking the mini-dozer to the crane. I had not been forewarned that this might happen, and was knocked down onto the bulldozer's seat. As I attempted to get up and try again, an Army sergeant jumped up onto the dozer and hooked the cable to the Skycrane. He'd clearly done this before. We both then jumped from the dozer, and away it went back to Camp Haines. When the loud helicopter noise abated, he said, "Don't let it worry you. Hell, that happens to everyone the first time." My Huey soon arrived to take me back to camp. I never again volunteered for mini-dozer duty.

A couple of the other guys in the battalion had their own unexpected encounter with a vehicle. It happened with one of the two brand-spanking-new concrete mixers mounted on International truck frames we'd brought with us to Vietnam. The trucks were ugly, but at least they were new, and no construction battalion would have been able to construct much of anything without concrete. Anyway, during one of the daily rainstorms—and it rained every single day for forty days—one of the mixer drivers left our yard with a load of concrete. He drove no more than two hundred yards when his truck slid off the slick dirt road and became stuck in a shallow drainage ditch. He was lucky it didn't turn over, and he left it running so the concrete would continue mixing while he walked a short distance to our maintenance shop. A friend of his was a mechanic there and the driver asked him to use a bulldozer to pull his mixer out of the ditch. The mechanic wasn't an experienced dozer operator, but he agreed to help. They secured several heavy chains to the dozer, and the mixer driver bummed a ride by perching on the dozer's fuel tank as the mechanic drove to them to the concrete mixer.

Surveying the muddy conditions at the site, the mechanic suggested using the dozer's blade to push the mixer backward, rather than pull it, out of the ditch. The mixer driver also thought this was a better plan since he wouldn't have to crawl into the mud to hook the chains to the rear of the mixer. But their attempt failed, and not only did the mixer stay put, they inadvertently pushed the mixer's power takeoff (PTO) shaft through the radiator and into the fan, shutting the motor off, which in turn, stopped the mixer from turning. In the ninety-degree heat, the concrete hardened pretty damned quickly.

The accident happened on a busy road, so it didn't take long before "those in authority" were on the scene. Both the mixer driver and the mechanic were instructed to use the chains—the ones they should have used to begin with—to pull the mixer backward out of the ditch, and then ordered to tow the mixer down to the maintenance shop and park it beside a large, diesel-powered air compressor. They soon learned the air compressor would power the jackhammers they would both be using to break up the concrete.

Over the next week to ten days, those poor fellows spent their days inside the steel mixer drum. After they removed the cover from the man-sized hole on the outside of the drum, which all mixers have so they can be cleaned, they would lower a jackhammer into the hole and start working at it. They'd rotate shifts, and as the jackhammer worked, cement dust would boil from the manhole. After the "inside man" had worked for a while, he would climb out to clear his eyes and lungs, and they'd switch turns. Each time they crawled out of the manhole, for a break, a cigarette, or a meal, they had to dust themselves off. They looked like dogs shaking themselves after having fallen into a pit of baking flour.

We could gauge their daily progress by watching the pile of concrete chunks grow when they tossed them out of the manhole. When they finally finished the job of jackhammering and removing the hardened concrete from the mixer, the pile looked more like two loads of concrete. Thank God this life lesson—to not use a dozer to try to push a concrete mixer out of the mud—was learned because of someone else's mistake.

Not far from where the jackhammer operators were working was the location of one of the most preposterous situations I ever witnessed. When our battalion arrived at Camp Haines, we took ownership responsibility of all equipment left by the previous battalion. A road grader, or a motor grader, is a construction machine with a long blade used to create a flat surface during the grading process, and an inventory check showed that our battalion was now in possession of a Galion motor grader, although it had not been included on any previous inventory list. It was so strange that calls were made and lists rechecked, yet no one found any record that the grader ever belonged to any battalion. In fact, no record was found that the motor grader ever existed.

Perhaps because of the oddness of the whole situation, our division headquarters sent word that we weren't allowed to use the motor grader and needed to get rid of it. This seemed so illogical that my fellow Seabees and I agreed that there must have been a miscommunication. Surely, the leaders of our battalion misunderstood, and we would soon be allowed to use the Galion. The grader operators said it was the best grader in our whole inventory.

But no. We soon learned that "get rid of it" meant exactly that, so two groups vied for the prize. The builders of our battalion requested that they be allowed to remove the engine, so they could reconfigure it as a generator to provide electricity to a local school. Their request was denied, as they were told, "If one school has one, then all the schools will want one." Once again, division headquarters sent word to get rid of the Galion motor grader. The other group, the mechanics, thought it would be sporting to have a contest to see who could guess the correct number of minutes the grader's engine would run if it had no oil. They drained the oil from the motor and removed the oil filters, and then fired it up and let it run wide open, at full throttle. I think it ran between seven to ten minutes before it rattled its last gasping breath.

After, the mechanics used cutting torches to cut the grader in half and a dozer pushed the Galion into the burial pit its operator had dug. It was then covered, and out of sight. The destruction and burial of an excellent machine was absurd, but nothing compared to the absurdity that came two days later when division headquarters sent word for us to remove the engine so the builders could install it at the aforementioned schoolhouse. Hey, at least the mechanics were excited to have learned the engine could run without oil for up to ten minutes.

Also preposterous: the used pieces of equipment were loaded onto barges and pushed off into the sea. We were told this was cheaper than sending all of it back home, but we should have just left the used equipment for the North Vietnamese to use after Saigon fell.

Two weeks before we left Vietnam, a dozen of us were sent twenty miles north of Camp Haines to complete a road project the previous battalion had not had time to finish. We were housed at an Army firebase, and when dinnertime rolled around on our first night, we only found one chow hall, the Army officers' mess. We entered, filled our plates, and enjoyed a hot meal while the Army officers glared at us, wondering why we were eating with them. When we left in search of a beer, we noticed the officers' mess was well lit. Its electricity was being provided by one of our battalion's generators, and our MCB (US Navy Mobile Construction Battalion) emblems affixed to both sides looked awfully good. This made us proud!

The next morning, we were not so politely ordered to stay out of the officer's mess. We were enlisted men, not officers, so we were told to eat C rations just like the Army's enlisted men did. We begrudgingly ate our cold cans of ham and eggs and then headed to the jobsite for the day. For dinner, we were suffering through more C rations when, much to our surprise, an Army officer stopped by our hut and implored us to join them in their mess for meals the remaining few days we were there. That same morning, we had been ordered to get out, but now, he wanted us back. What the hell was going on? Our master chief soon made us aware that he had hooked his jeep to the generator during the day and hauled it to an off-site location. The generator was the Army's sole source of electricity, and powered their lights, hot water, and, most importantly, the electric stoves in their kitchen. He said that as far as he was concerned, the Army must have misappropriated the generator, since it was clearly labeled as US Navy Seabee property. He informed us that he had rightfully reclaimed what surely belonged to us. This master chief was one of the older recruits who had been instructed by a judge to either join up or go to jail. He was an extremely likeable fellow, and even though he was our boss since his rate was an E-7 and most of us were E-4s or E-5s, he hardly ever cared enough to make decisions. For every nine out of ten questions he was asked, he answered, "Not knowing, I would hesitate to say," or a similar statement. Truthfully, stealing the generator was probably his only worthwhile act while he was in South Vietnam.

After we accepted the Army officer's invitation to dine with them, our master chief towed the generator back to the mess hall and reconnected all the wires. We enjoyed hot breakfasts and dinners for the remainder of our stay at the firebase just south of Quảng Trị, and the officers realized that we weren't such bad guys, even though we were lowly enlisted men. It really is surprising how quickly folks can change their minds when you turn their lights out.

While finishing the previous battalion's work, our small detachment was assigned several tractors and lowboys, which we used each morning to haul our equipment to the jobsite. We were working in gleaming white sand and the temperature was over one hundred degrees each day. Throughout the day, rain would fall; the sun would come out and cause steam to rise from the sand; it would rain some more; and then more sun and more steam. The weather made us feel like we were in the middle of a gigantic sauna; teak benches would have completed it.

Fortunately, the girls at the convoy point sixty miles away must have had sisters outside Quảng Trị who also knew the drinking habits of American soldiers, and since the temperature required that we stay hydrated, we were more than happy to purchase a few ice-cold beers from them. By the time lunchtime rolled around each day, we had been drinking those soothing beverages for six or seven hours straight. Although our work was physically exerting, we felt no pain thanks to the ice-cold beer.

We were required to keep our weapons with us at all times, but one of the guys, Bob, the friend we'd pulled out of his locker during the mortar attack, figured his M16 would stay much cleaner if he left it in his parked truck rather than having it ride with him on his bulldozer all day. This worked okay for him until we returned to our trucks for lunch one day and he noticed his weapon was no longer there. What a fix he was in—high as a Georgia pine from hours of drinking and now missing his M16.

Earlier that morning, one of our guys observed the usual number of kids around the site, including our angelic beer servers. Now, at lunch, all the boys had hightailed it and were nowhere to be seen. Thankfully, our beer girls stayed on duty.

Bob asked for my help, so we composed ourselves enough to act like badasses, although two shirtless and extremely suntanned guys with Colt 45s resembled armed surfers more than anything else. Beer dictated that our best course of action was to march right into the middle of the closest village and demand the gun back. We knew maybe five or six Vietnamese words or expressions, such as *di di mau* (meaning either go quickly, or get the hell away from me); *beaucoup*, pronounced "bookoo" (from the French word meaning a lot, or many); *numma ten* (from our English "number ten," which they said to mean you're a cool dude); *numma ten thou* (from our English "number ten thousand," which qualifies you to be a piece of shit, and that you can go slam to hell); and *boom boom?* (meaning "Would you like to have sex?"), so we were sure they'd understand us. Bob and I walked about two miles down a well-worn path to find a small welcoming committee. Somehow, we communicated our need to speak with the village elder, and sure enough, we were led to Papa San's hut.

By this point, what we really needed were a few more beers, but we accepted Papa San's (repeated) offers for hot tea, hoping we weren't about to be poisoned. The tea was poured into regular drinking glasses, and Bob and I watched strange leaves swirling in the liquid. Realizing we had to die someday anyway, we downed the surprisingly tasty glasses of tea, although we declined his offer of refills.

We painstakingly explained Bob's missing M16 predicament, including the fact of the young boys' presence at our jobsite. We spoke slowly, enunciating each syllable, and we thought we might be getting our message across to him. All of a sudden, Papa San scared the shit out of us when, quite unexpectedly, he loudly screamed something at another Vietnamese man sitting barely three feet from us in the hut. That man jumped up, ran outside, and began emphatically yelling. Other villagers then started yelling too, and all Bob and I could do was wonder what the hell was going on and what we had done to cause their strange behavior. We figured we were done for and were about to die, but, for some reason, most likely alcohol, these outbursts struck us as just-plumb hilarious, and we damned nearly choked trying to keep from laughing in the old man's face. Our laughter came close to causing us to spit out our mouthfuls of excellent tea.

While the villagers were yelling, Papa San rose and informed us he would do all he could to find the gun, or at least we think that's what he said. We offered our thanks by bowing and shaking hands, and then we either walked or staggered back to where the trucks were parked. During our walk, each time we looked at each other, we broke into uncontrollable laughter. Fifteen minutes later, as we related the story to our buddies and were all cackling heartily, Papa San walked up the path to where we stood, accompanied by a sullen kid. He asked, we think, if Bob's truck had been parked in the same location all day. We assured him that it had not been moved and then watched as he walked over to the truck and stood by the passenger-side running board. From there, he took four or five steps directly away from the truck, and then turned left ninety degrees and took a few more steps. There, he stopped, bent down, and dug in the sand a few inches, and voila—the M16! The kid was looking down and wouldn't make eye contact with anyone. I'm sure he was thinking of the punishment he'd receive on his return to the village.

Without a doubt, Bob and I were certain of the fact that drinking beer had helped us recover his M16, so we thanked Papa San by giving him a six-pack of cold ones, plus all the C rations we had left. I'll bet the little culprit's arms were sore by the time he carried those heavy items back to his village, because Papa San loaded his ass down.

On our last night at the Army camp south of Quảng Trị, our detachment of twelve thought it only fitting to celebrate going home by having a few beers while we rehashed some of our "in country" war stories. Our few beers turned into many beers, our war stories grew in size, and soon we were reveling. Our celebration ended up lasting the entire night. Just about the time the sun was coming up, someone hollered, "Damn, look at Wally!"

We gathered by Wally's bed as he lay flat on his back. By the looks of it, he had, for quite a long while, been placing his empty beer cans on his chest, and the large number of cans indicated that others had added their own empties. He was bare-chested, owing to it being in hotter-than-hell Vietnam, and his entire chest, stomach, and parts of his thighs were covered by empty cans.

We began applauding Wally's efforts when the door of our hooch suddenly opened. Our commanding officer strode in, followed closely by his executive officer, and then by more officers than I could count. We just knew our CO had come to thank us for completing the road project on time, and although I don't remember his exact words after he and the executive officer stopped at the foot of Wally's bunk, I do remember they were directed at all of us and they were far from complimentary. But to Wally's credit, he had quite a skill for balancing beer cans (or maybe he was too inebriated to move). He held perfectly still while the CO spoke and didn't let a single can fall from where it had been placed.

The original orders for that morning had been to pack up and leave Quảng Trị, and take all our equipment back to our base camp. To our dismay, we realized just how badly we had screwed up when we were ordered to "go out there and act like it is just another work day." Our punishment, even though none of us were anywhere close to sober, was to operate our heavy equipment in the 115-degree heat. The heat had somehow caused some of us to fuck up work we had thought we'd already completed.

I spent all morning sweating and wanting to hurl. But orders being orders, we did our best. In the early afternoon, probably timed to coincide with when everyone was nearing sobriety, we received word to head 'em up and move it out, which meant that we could gather our belongings, including all our earthmoving equipment, and go back to Camp Haines.

Back at the camp, we found our 850 fellow Seabees reveling in the awesomeness of leaving Vietnam the next day. They were having such a good time because the CO hadn't found them shit-faced at daybreak that morning and they hadn't been ordered to act like it was just another work day. Instead, they'd spent the day packing their belongings, and various types of contraband, such as M16s, Colt 45s, bayonets, and anything else they could find, into duffle bags and footlockers. The twelve of us coming from Quảng Trị—now sober, but very hungover—were told that we had already celebrated, so we needed to use what time we had left in Vietnam to pack our shit.

I chuckle every time I remember the look on the CO's face as he stood at the foot of Wally's bunk, or the look on our executive officer's face as he looked down at all those empty beer cans perched on Wally's chest. I could swear his suppressed grin said, "Damn, I wish I could do that."

Most veterans don't talk much about their war experiences, and I have mostly been the same way since I left active military service at the end of 1969. But in recent years, memories have returned to me that I have not dwelled on for a very long time—most of them lighthearted, humorous, and alcohol-infused. I do recall difficult times, and certain sounds stick in my head: the *thump, thump, thump, thump* made as B-52s dropped their bombs along the Hồ Chí Minh trail in the A Shau Valley, twenty miles west of our camp, and the explosions that came from the large craters they blasted, and the *whoosh, whoosh, whoosh* from sixteen-inch shells fired from the USS *New Jersey* as they flew over Da Nang. Maybe I have suppressed the worst memories for so long that I can no longer access them. I honestly don't know.

Of one thing I have crystal clarity: my memories of Vietnam are much different than those of the brave soldiers who had to fight and kill every day; who traipsed through, and slept in, snake-, scorpion-, and leech-infested, Agent Orange-sprayed jungles. Every time I think of the Army soldiers we saw in the Da Nang airport—those possessing the thousand-yard stare—I am very thankful that I have never suffered the nightmares those soldiers probably had, or still have. Nor have I suffered the PSTD symptoms, or the Agent Orange-related health issues that many of those who served in Southeast Asia had. I thank all of them for their service, and I'm grateful I did not have to walk in their shoes.

I admit that many of my memories were either caused, or made much more hilarious, by the enjoyment of alcohol, mostly beer. Without the beverages, what stories would I have? "Arrived at the convoy point, had to sit for five hours with nothing to do, read some, and smoked cigarettes"? I'm not really sure why, but the group of guys I worked with and called friends never smoked pot, otherwise I could have sat at the convoy point in the middle of a purple haze. Lots of the other guys smoked what they claimed to be the best shit they'd ever had, and maybe it was. I suppose drinking was good enough for us. Whatever it takes to get you through.

To be honest, we probably drank so we would not be plumb scared to death of being killed. If I had become an Army ground-pounder after being drafted, I think I could have handled actual combat, but waiting around to be shot at or hit by mortars, or step on a land mine, just left too much to contemplate and worry about. And there were some good perks to being a Seabee. Besides the amazing food we had at our camp, we also were able to enjoy Da Nang's My Khe Beach, which we called China Beach, from time to time. It had beautiful white sand and was a welcome escape from reality where men and women GIs sunned themselves and cooled off in the South China Sea. It didn't matter that the beach was ringed with barbed wire, it still gave you a feeling of being somewhere much more serene than war-torn Vietnam.

Being a truck driver also helped make my experience tolerable since it gave me a unique vantage point from which to observe Vietnam's sights and culture. I drove through the same villages and towns several times each week, including Huế, the Cathedral City. I saw new construction by villagers, some of whom used pilfered telephone poles for planks or made siding from cardboard C ration boxes. I got to see South Vietnamese people more than some of my fellow soldiers, and they struck me as being mellow and kind. The women, many of whom were quite beautiful, wore white tops, black pants, and large straw or bamboo sun hats. Most either walked or rode bicycles everywhere they went. Many older women chewed betel nuts to achieve a slight buzz and a warming sensation in the body. It wasn't hard to spot a betel nut chewer; their teeth were dyed red and they were always spitting out the scarlet juices. Small, elderly women could hold more weight on their shoulders than most of the big GIs, sometimes carrying two four-foot-tall bundles of roots. They could also haul two pigs or two baskets full of chickens.

The South Vietnamese kids were curious. They wanted to learn about us Americans, and always smiled at us. They walked, or ran, everywhere on their bare feet, and picked up everything GIs left behind to take home to their families, just knowing something useful would become of these items. The young boys rode water buffaloes as the animals pulled plows through rice paddies. The young girls served us ice-cold beer, and never ran out.

Some Vietnamese military men walked hand-in-hand as they strolled along the roads. Some were cruel, pushing civilians out of their way as they strutted along, and I hardly ever saw them smile. But I doubt I would smile if my country had been at war since

111 BC, when China conquered what is now the northern part of Vietnam.

Smoke was everywhere, ever present, and charcoal choked the air from the burning of various roots. At daybreak each day, you would see the bare asses of hundreds of Vietnamese people of all ages, as they bent over to use the Perfume River, which crosses through Huế, as a giant outhouse. At dawn one day, on a sampan on a tributary of the Perfume River, I saw a woman relieve herself off the leeward side, then draw a pan full of water from the wayward side to use for fixing the family's breakfast rice.

I saw all this and more from up on my trucks, and I also dealt with Vietnamese vehicles. Buses resembling Volkswagen campers, designed to carry maybe twenty passengers, carried at least sixty adults plus the passengers' kids, chickens, and pigs. The bus drivers seemed to only utilize one speed, full speed ahead, and the riders would cling to the sides of the busses as they flew down the road. If the busses weren't flying, they were stopped. Upon approaching the rear of one of these buses, while driving a truck, which seldom happened because the drivers all drove like bats out of hell, I was sure to keep a considerable distance behind it, ever mindful that one of the passengers sitting on the bus's roof could fall onto the roadway.

One of my great memories from Vietnam is of the time I had an unexpected run-in. I was holding the screen door of our chow hall open for an Army soldier a few steps behind me. The guy was hunched over because he was wearing a heavy flak jacket, while carrying his M16, a Colt 45, an M79 grenade launcher, an enormous backpack full of who-knows-what, and several web belts full of various types of ammunition, including hand grenades. But he wasn't just any

Army guy, he was a friend who grew up a half mile from my childhood home. We were thirteen thousand miles away from home, and somehow happened to cross paths. He had come out of the jungle for a few days and was very glad to eat something other than C rations. We ate together and reminisced about growing up on Airport Road. He made a point of telling me how lucky I was to be able to eat in my chow hall every day instead of eating out of a can while sitting in the jungle in the ever-present rain. Despite everything around us, being together was a reminder of home. For just a moment my mind flashed back to Obie's farms and tractors; but after my friend left it was time to head out to my flatbed filled with highway-building supplies.

Our battalion's tour of South Vietnam was eight months long. When we arrived back to the chaotic USA, we were greeted in San Francisco by Hare Krishna chanters calling for world peace, and some hippies who spat at us. We were granted thirty days of leave and most everyone headed home to see his family. After leave, we were to spend the next six months in California, preparing for another eight-month deployment to South Vietnam.

Six weeks prior to our redeployment, we were notified of a troop reduction directive, which stated, among other things, that if the date of our release from active duty fell on or before a certain date, then we were immediately released from active duty. I made the cut-off date by ten days. I had signed on the dotted line to serve in the Seabees for two-and-a-half years, but thanks to this early out, I spent only twenty months and nineteen days on active duty.

While in California, I lived in an apartment with another North Carolina fellow, who also made the early out cut-off date. On

December 19, 1969, we loaded my motorcycle and his guitars and amps into the trailer he towed behind his car, and headed home to enjoy the rest of our lives.

PART THREE

COLLEGE TRUCKING

Before enlisting in the Seabees, I attended college for one year. Upon release from active military duty in 1969, I returned to East Carolina University located in North Carolina. Over the next several years, each time my veteran's benefits money ran short, which unfortunately happened pretty often, I would leave school for three months, a quarter of an academic year, to either drive tractor trailers or operate heavy equipment. I would save most of what I earned, and then return to classes until poverty dictated that I needed to work again.

At one point I decided that I needed to take a few months off to make money. An opportunity arose when my high school girl-friend's brother-in-law (who owned a marina thirty miles away from my college) and his father decided they wanted to open a trucking company. They needed drivers and knew my trucking background, so it seemed like a good fit.

Unfortunately for me, their company used old, worn-down trucking equipment, which came with many perils. The first trip I made for them was supposed to be relatively short, hauling a load of plywood from Eastern North Carolina to North Wilkesboro, North Carolina, but the equipment didn't hold up as it should have. When I was within thirty miles of North Wilkesboro, I looked out of my passenger-side mirror and saw heavy light blue smoke. I pulled over at a wide spot to figure out where it was coming from, and saw that the smoke was billowing from the right rear wheel of the tractor's tag axle. It's called a "tag axle" because it's not connected to the truck's driveshaft, so it has no pulling power and essentially just tags along.

When I climbed out of the cab, I took the fire extinguisher with me in case the wheel caught fire.

It's a good thing I had a wide spot to pull over on, because a wrecker needed to get close to the wheel on the right side. When the wrecker lifted the axle, the entire wheel assembly, brake drum and all, fell off and rolled into a ditch, as the wheel had broken off of the axle. Inspection showed that the bearings had become so hot that they were fused onto the axle. While the wrecker held the axle up, I chained it to the tractor's frame. When the wrecker lowered the axle, it returned to the height it would have been if the wheel had stayed on the axle. I then had the wrecker pull the wheel assembly out of the ditch and place it onto the rear of the trailer, where I also chained it down.

With the axle more or less able to do its job again, albeit missing a wheel, I slowly traveled the rest of the way to North Wilkesboro. When I arrived at the building supply store, the drop-off spot, it was clear that the employees and customers must have never before seen a wheel-less tractor with a chained-up axle. Everyone in the building came out back to take a look. On my way back to Eastern North Carolina, I learned that the enforcement community doesn't have a problem with a chained-up axle, as I was waved through the weigh station on the northbound side of I-85, just south of Durham, North Carolina.

The next morning, when I spoke with the father, it became pretty clear to me that I wouldn't work for these fellows for very long. He told me the bearings had run out of grease, which had caused them to get very hot, due to my incomplete pre-trip inspection the day before. I objected, but figured it would be pointless to argue with his incorrect assessment. The assessment of the mechanical problem was correct, but the claim that I'd done an incomplete inspection was incredulous. A driver's pre-trip inspection does not include a mechanic's job of pulling a wheel to inspect that the bearings have sufficient grease.

I ran another one or two trips for these guys before I quit working for them, and within a week after I quit, I went to work for a construction company. Having operated scrapers, better known as "pans" in Vietnam, I spent the next two months running a scraper—a tractor used for digging, moving, and leveling ground—on the property of a chemical company. I would load between eighteen and twenty cubic yards of red clay at a borrow pit, haul it over a dirt road for close to a mile, and then dump each load on the top side of a deep ravine. As I would head for another load, another employee would operate a bulldozer to push this red clay over into the ravine. We were covering up fifty-five-gallon drums of some unknown-to-me chemicals that a company had dumped. I was skeptical about what would happen if, and when, the drums rusted open, but I didn't raise my concerns. I was getting a paycheck. Not too many years later, there was a pollution contamination catastrophe at the Love Canal neighborhood landfill in upstate New York. Hundreds of residents were sickened and the event culminated in an extensive cleanup operation by the Environmental Protection Agency's Superfund Task Force. This made me wonder if I had been a party to that same kind of environmental tragedy, and, evidently, I still think about it today.

After another few months at college, I began working for WMTS Trucking. WMTS's management was okay with me driving for a few months at a time and then returning to college in between. They didn't remove my driver files when I went back to school, so periodically, I would also run short trips on weekends. I was very fortunate to have found such a decent group of trucking folks for work for. After forty-five years, I am honored to still have them as friends.

The company was both a flatbed and closed van operation. I mostly stuck with flatbeds, and I tip my hat to all truck drivers who pull flatbed trailers. The work is hard enough in good weather, and worse when it's hot, cold, raining, or snowing. Many times, after getting filthy while tarping and securing loads, drivers have to change into clean, dry clothes before getting behind the wheel. My limited experience pulling flatbeds taught me to respect their work. The adage "Someone's gotta do it" does not fit the flatbedder's mentality. They do it because they are good at their jobs, and they enjoy getting a job done well.

WMTS hauled an awful lot of lumber, and most of those loads were both easy to load and relatively easy for the driver to chain and tarp. Having previously had my ass chewed out royally by the Army colonel, I never left a shipper without making damned sure the load was properly secured. If there were enough chains, binders, and nylon straps available, they were all put to use for holding the load in place.

An example of an easy load would be a truckload of pressure-treated lumber. These loads do not require tarps, so the driver uses six to eight nylon straps and then he is on his way. Of course, there is always that shipper who causes you to bitch and moan when you get dispatched to its place of business. For me, it was a lumber mill that shipped finished tongue-and-groove lumber, which is as slippery as hot molasses, or other adjectives I will refrain from using.

Carrying my first load of this shit—what we drivers affectionately called it—was an educational experience. To begin with, after the forklift placed all the various sized bundles on my trailer, it looked to me to be over the legal height limit of 13′6″. I was sure the load was too high, so I asked the forklift driver if he could measure it, and was disheartened to see that it measured exactly 13′6″ from the ground. I let out a long inward groan for being wrong, while the fellow who'd measured just right was grinning.

If I remember correctly, the flatbed chains that had one end welded to the rub rail had half-inch links. A trailer's rub rails protect the vehicle in the event it rubs against something, like a barrier or a wall. They're important assets when carrying a load because they protect the chains, straps, and ropes that secure it. One end of each of the eight chains was welded so they would not be stolen, or slide off the trailer. Due to the chains being too heavy to throw up and over a tall load, the only way to get the chains across it was to throw a rope over the lumber from the driver side to the passenger side of the trailer. On the passenger side, you would tie the rope to the end of a chain, and then on the driver side, you would pull the rope-attached chain across the load of lumber. Then, all you had to do was repeat this process seven more times, for each of the other chains. Before hooking the chain binders, the driver had to climb on top of the load in order to place corner protectors under each chain, otherwise, the chain would bite into and damage the lumber.

For my first load from the shipper of the finished tongue-and-groove lumber, I wanted to be extra cautious. I reckoned that tighter chains always worked best in securing a load, so I used a three-foot-long pipe, known as a binder pipe, on the chain binders to make sure the chains were secure. When I was finished, by God, those chains were tight. The next step was to tarp this expensive load of flooring. WMTS's canvas tarps had ropes threaded through steel grommets (eyelets) spaced every twenty-four inches. These grommets protected the tarp from being torn. After rolling out the tarp on top of the lumber, I would pull the sides down until they were equal on both sides of the trailer. Using the ropes, I would then begin the process of cinching each one by tying knots to the rub rails on both sides of the trailer. I could only hit the road after completing this whole routine,

These ropes led to an especially good time when you arrived at your delivery destination after driving through snow and ice, and found that your carefully hand-tied knots were frozen solid. Creativity was your best friend when dealing with frozen knots, and the best way to untie them was to find a way to thaw the knots. One time I used the receiver's bathroom to fill gallon jugs with hot water to douse the knots. It took quite a few trips, getting water and dowsing, and repeating, but it got the job done. Another time, when delivering to a building supply store, I had to thaw the knots by purchasing a small propane torch. On one occasion, my only solution was to cut every single rope. Thankfully, flatbed trucking has come a long way since the days of the canvas tarps. Ropes have been replaced by rubber bungee cords, and today, many flatbed trailers employ a tarp attached to an aluminum structure, which can be rolled forward from the rear to the front, thereby opening the trailer bed for loading and unloading.

After leaving the tongue-and-groove lumber mill, I traveled about one hundred miles before I noticed in my mirrors what seemed to be bulges under the tarps on each side of the trailer. I found a place to pull over to inspect the load, got out, and saw that the lumber bundles had worked outward and were bulging out from under the tarp. The load looked pregnant. I had tightened the chains so tightly that the bundles were actually flattening on the top, while at the same time being squeezed out at the sides. I learned a life lesson that day: When hauling finished tongue-and-groove lumber, only tighten chains so there is no play in them. Do NOT tighten the shit out of them!

Listening to other drivers' stories proved I was not the only one who hated hauling finished lumber. It was so ornery to transport that sometimes, even though you were sure the load was securely tightened, the lumber would seem to develop a life of its own. While trucking down the highway, individual lumber pieces would begin working their way out of the middle of the top bundles on the back of the trailer. These rogue pieces would "telescope" so much that they would poke holes in the canvas tarps.

I was a rookie at hauling finished lumber, so I was certain several drivers were pulling my leg when they said there was a truck stop north of Richmond, Virginia, that had an outside brick wall on one side of its repair shop, which everyone used when they hauled finished lumber. I thought, *Yeah, right! What idiot would back their truck into a brick wall?* But when this first load did, in fact, telescope, I learned they weren't kidding. I pulled into the truck stop and then slowly backed my trailer up against its brick wall, thereby pushing the offending telescoped pieces back into their bundles. I was thankful at that moment, but it didn't change my feeling overall. I'm sure older flatbed operators would agree that slick, finished lumber was among the top ten most aggravating commodities to transport.

I was never overjoyed to learn that it was my turn to be dispatched to pick up a load from this lumber shipper, but I never again had as much trouble as I did with that inaugural load. I even learned the trick of doubling the tarp over the end of the load, which helped create more of a barrier that kept the lumber from telescoping. There were times, however, that even that didn't do the trick.

Another life lesson learned during a hiatus from college was that traffic circles have posted speed limits to alert drivers that they need to slow down for good reason. One time while pulling a closed van and hauling a palletized load of cut paper stacked on pallets with four-inch-high "feet" (small blocks of wood that raise the height of the pallet), I entered the only traffic circle in Newton Grove, North Carolina. (At the time, it may have been the only traffic circle in North Carolina, at least that I was aware of.) I was clearly going too fast: while rounding the circle, I checked my driver's side mirror to make sure I was missing the curb, and, lo and behold, I saw that my driver's-side trailer wheels had come off the ground.

I am positive I yelled either, "Oh, shit!" or "Oh, fuck!" and I am certain that shouting these words caused the trailer wheels to sit back down on the pavement. After successfully making it through the circle and pulling over to the side of the highway, I was horrified to see that the front of my trailer was tilting over and damned nearly sitting on the passenger-side tractor tires. I opened one of the trailer doors and I'm sure all sorts of cusswords were released since all the pallets of paper were sitting against the right-hand wall. All the four-inch feet, which I had never seen used before, had broken off when the pallets shifted. I badly wished I had observed those speed limit signs, because the next several days proved awfully damned agonizing.

The next morning, I asked some local folks if there was anywhere in town that regularly received shipments from tractor trailers, and one of them recommended a certain feedstore, which had everything I needed: a loading dock, available warehouse space, and a decent forklift. The very nice and helpful proprietor said he didn't have time to help me, but I was welcome to use the facilities and equipment on my own. It took most of the day for me to rework the load. I had to unload each pallet of paper, and due to the fact that the feet were broken all to hell, I had to raise each pallet high enough that I could use both a claw hammer and a pry bar to remove the broken pieces from under each pallet. I also made sure I didn't place my body too far under the pallets, just in case the forklift dropped, which would have smashed me flat. I reloaded the pallets after completing the foot surgery, and secured the load from shifting by nailing two-by-fours beside each pallet. Thankfully, I could do this because the trailer had a wooden floor.

The following morning, I arrived at the printing plant in Manhattan. In the way that only a true New Yorker could, the receiver greeted me by saying, "What the fuck is wrong with that fucking paper mill? These fucking pallets were supposed to have fucking four-inch feet on the bottoms, so we can unload the fucking things with our pallet jack. We can't get our fucking jack under these fucking pallets! What a way to start the fucking day!"

After I told him what happened, he said, "Fucking truck drivers!"

Of course, he refused the load because the pallets had no feet. Several hours later, I delivered the fucking load to a warehouse in Jersey fucking City. This fucking warehouse installed new four-inch feet on each pallet and then redelivered the load to the paper plant in the city. I blamed everything on that fucking traffic circle.

On another break from college, I returned to WMTS to drive for three more months. This is when I met Junnie Jones. Junnie can't be put in a box. He certainly danced to the beat of a different drummer and was either one-of-a-kind, or the mold was thrown away when he was born. I'm sure as hell that I will never meet another person like him.

For close to a week, Junnie and I received duplicate dispatches. Each of our loads were picked up at the same shippers and delivered to the same consignees. We both questioned the accuracy of each dispatch, because it rarely happened that more than one load shipped daily from the same shippers to the same consignees, but sure enough, there were two loads waiting for us every time we pulled into a shipping location. We had several short runs of three hundred to four hundred miles and a couple longer trips of six hundred to seven hundred miles.

Due to the fact that we ran together, we stopped at the same truck stops for fuel and food. Usually, when you spend this much time with another person, you get a pretty good idea of what makes them tick; with Junnie, I never did. After several meals together, I noticed that he always let me be the first to order, and then told the waitresses he would have what I was having, including whatever I'd chosen to drink. It didn't matter what I ordered; Junnie always got the same. Curiosity eventually got the better of me, so I asked him about this. He said he didn't much like or dislike any foods, and that the only reason he ate anything was because he had to eat to stay alive. He felt it was a hell of a lot easier to copy someone else's order than to suffer the aggravating chore of having to make his own selections. If Junnie traveled alone, he would sit at the counter at truck stops and drink coffee until another driver arrived and sat beside him, and then he'd order what the other guy got. He said there were many times he had to laugh it off when one of the drivers referred to the "strange fellow" or "fruitcake" sitting next to him.

As Junnie and I talked more, my assumption that his eating habits were only one of the unusual things about him was confirmed. He told me a number of "out there" stories and a good bit of them involved shotguns. One of them was of an event that took place several years before I met him. He said he'd been living in a house in a swampy section of Eastern North Carolina, and one day watched a snake crawl under his house. He didn't think much about it at the time, but over the next few days, he saw several more snakes crawl under his house. When he finally opened the door to the crawl space, he found a den of snakes. Fuming, he went into his house, loaded his shotgun, reopened the crawl space door, and then, to borrow his words, "killed every one of those sons of bitches." Unfortunately, he also killed most of the plumbing and electrical wiring running throughout the crawl space. He said he replaced the pipes and wiring by himself, and that although it was pretty expensive, at least he didn't have to crawl around with live snakes down there. Kind of makes sense to me . . . kind of . . . .

This snake killing story sounded so bizarre that if anyone else had told it, I would have been sure it was made up. But if Junnie couldn't even think up his own food choices, I doubted that he could concoct his own stories, and damn if I wasn't spot on in that judgment. Friends later told me they'd read about Junnie and the snakes in their local newspaper.

Another of Junnie's stories was of him being awakened during the night to find his wife straddling him and pointing a loaded handgun at his face. He asked what the hell she was doing, and she said that she would blow his damned head off if he ever messed with another woman. Junnie asked why she was saying this in the middle of the night, and she informed him that she couldn't sleep; she needed to tell him how she felt. When Junnie finished this story, I just sat there shaking my head. For once, I was not able to utter a single word.

Another story Junnie told me was that one day while he was sitting at his kitchen table cleaning his shotgun, his wife called down to him from the top of the stairs to ask if he was watching the television, which was playing in the den. He answered that he'd been watching, but wasn't watching at that moment, so she asked him to turn it off. Half an hour later, she hollered from the top of the stairs, "I said to turn off the goddamned television!" At this, he calmly loaded his shotgun, pointed it at the television through the den doorway, and turned it off by blowing it all to hell. His wife ran down the steps yelling, "Why did you do that, you crazy bastard?!"

"Well, you told me to turn off the goddamned television," he replied.

"What if I had told you to turn off the oven or the dishwasher?" she asked.

Junnie must have taken it as a challenge.

After all the appliances were blown, the police hauled Junnie off to the pokey. At his court appearance, he told the judge he didn't know why everyone was so upset. Everything he had blown to hell was his, so what did it matter? The judge replied, "No, Mr. Jones, everything you did not shoot was yours. Everything you shot belonged to Mrs. Jones, and you're going to replace all of Mrs. Jones's appliances that you shot."

With a straight face, and in a manner he might have used to describe a mundane task like taking out the trash, Junnie told me he bought a new television, dishwasher, range, refrigerator, and washer and dryer. It seemed like it was just something that happened every day.

Junnie told me that story just as our week together was coming to an end and I was relieved about that because the story left me shaking my head so much that I got a terrible headache. It sounded so bizarre that I felt sure he had made that shit up. But damn was I wrong; friends told me they had also read about this one in their local newspaper.

For years after I left WMTS I wondered what happened to Junnie. All I knew was that he was there for a while and then went to drive for someone else, and I assumed that his shotgun must have continued to be a big part of his life. Just recently, I learned that he has driven for the same company for many years and is soon to retire. I believe he managed to take care of his demons, even if he probably blew the shit out of them.

———

We all hear horror stories of how badly cops have treated truckers, but I have no stories like this to tell because most of my experiences with the police were positive. One officer found me lost while roaming the streets of Manhattan one spring morning just after daylight. Observing one of New York City's finest sitting in a three-wheeled cop mobile—which resembled a golf cart with NYC police markings—parked at a curb, I stopped and asked him for directions to a printing plant. He told me I had driven past the address, and explained that I needed to turn around and go back the other way.

He said, "Can you swing it around in this intersection?"

I replied that although there was enough room to make a U-turn, the traffic was too heavy to attempt the maneuver.

He looked at me and with a wink in his voice asked, "You think so?" Then, to my surprise, he climbed out of his three-wheeler, walked into the middle of the intersection, blew his whistle to indicate that all traffic, from all four directions, should stop immediately, and looked at me and shrugged his shoulders, as if to say, "Well?"

I quickly performed my 180-degree turn, waved thanks to him, and took off down the street. Half a block later, I heard his sirens and saw him approaching. I knew he wanted me to pull over and as I did, I was pissed. I thought, *Damn, he set me up so he could give me a ticket for making a U-turn.* But I'd thought too little of him, because far from ticketing me, he walked up to my vehicle and said he'd remembered the delivery address as being a "motherfucker" for truckers to get to. It apparently required that truckers back into the dock from their blind side. In trucking terms, the blind side refers to the passenger side of the vehicle, where the view can only be observed by looking through mirrors, as opposed to being able to also looking out the driver's side window while backing up. He told me to follow him and not only did he lead me the half-dozen blocks to the correct address, but once again he got out of his vehicle to stop traffic for me. Any driver who has said NYC cops were assholes hadn't met this fine gentleman.

Once in Chicago, I was directed to pull into a delivery alley to unload some furniture at a store downtown. I broke down four or five widow-maker heavy desks while the two lazy, bastard receiving clerks watched me damned nearly kill myself. When I finished unloading, I climbed into the tractor to continue through the alley since the receiver had said every other trucker did this, but, dammit, evidently the other trucks were smaller, because my 13′6″ high trailer wouldn't even come close to clearing some of the fire escape ladders protruding down the length of the alley. Since I couldn't proceed without knocking them over, I slowly began backing up toward where I'd entered the alley, but I had to stop when I reached a busy sidewalk and an even busier roadway.

I have no idea where he came from, but one of Chicago's finest noticed my dilemma. He didn't even ask me if I needed help, he just went ahead and stopped oncoming traffic by parking his car to block the lanes. The policeman got out of his car, halted the sidewalk pedestrians, and then directed me to continue backing onto the highway.

Another good experience was with a Maryland state trooper on US Highway 301 North. Close to midnight, as I crossed the William Lane Chesapeake Bay Bridge east of Annapolis, CB-equipped drivers informed everyone that there was a "smokey," our word for a cop, entering 301 North from Highway 50. I was driving the speed limit while pulling a refrigerated load of pharmaceuticals, so I maintained my speed knowing the trooper was behind me.

The trooper soon passed me, and I flashed my lights to indicate to him that he'd safely passed me and could come back into my lane if he wanted to. He did, and he came on the CB and said, "Thanks, Mr. WMTS." We conversed for a few minutes and he asked where I was headed. Replying that I was staying on 301 North, while heading to New Jersey, he told me he would not have a problem if I wanted to pick up my speed if the truck was capable, since he was the only trooper out tonight.

It didn't take too long before I started creeping up on him, and he said something like, "Man, I guess it is capable of more than the speed limit!"

The trooper and I ended up having about ten miles worth of good conversation. When he had to exit, his parting words were, "I just checked and I'm the only cop out here tonight. The highway is clear to the Delaware line, so you can keep the hammer down. Hope to talk to you again someday."

I never saw, or heard, from him again, but the nice memory is still vivid.

For the most part, very few truck drivers enjoy runs into New York City or Long Island. The exception to this are runs for deliveries being made near the end of Long Island, where the scenery is spectacular and the real estate makes you wonder where all that money comes from. Unfortunately, most truckload freight deliveries occur within about ten miles of crossing one of the river bridges. It's also often necessary to deal with traffic at rush hour, which is always bad to horrible. Friday night rush hour, especially during the summer, is the worst, since everyone is in a hurry to get to the Jersey Shore.

I had the misfortune of having to deliver to Maspeth, Queens, late one Friday afternoon in summer. The crush of cars, trucks, buses, and recreational vehicles clogged the roads so badly that traffic was inching along at just a few miles per hour. As I passed an on-ramp, and with no warning, I watched as the front end of a car tried to sneak between the front bumper of my International and the rear bumper of the car ahead of me, even though there must have been less than two feet separating us. Then came a crunching noise, along with a pulling motion on my steering wheel. Thankfully, there was an exit within one hundred yards. I pulled onto the exit so all the other drivers wouldn't be cussing any more than they were already.

Thanks to Lady Luck, there was also a phone booth at the end of the nearby off-ramp, so I pulled over, got out of the tractor, and headed for the phone—which is when I saw the driver who'd tried to squeeze in front of me pull into the off-ramp and then park in front of my truck. He jumped out and began yelling and pointing to his car, "Look where you tore up my boss's station wagon!" He handed me his license and registration and I wrote the information on a piece of paper and gave them back to him. He told me to hand him my license and registration too, and I assured him that I would do so when the police arrived. When I turned to dial 911, the driver hollered at me, "You aren't going to give me your fucking information, are you?"

Once again I informed him that I would give my documents to him when the cops got there. Still yelling, he called out, "Fuck you, asshole!"

He jumped in the car and squealed the tires as he roared off. As he was leaving, I verified that the license plate number matched the registration he had given me.

After I called 911 and reported the incident, I asked if a cop could come to write a report. While I was waiting, I noticed some oil leaking from the plastic bearing hub cover on the front of my tractor's right wheel. It had been broken and punctured when the station wagon hit it. I wiped it as clean as I could, and then learned another of the one thousand uses for duct tape. (Impressively, the tape would stay in place, preventing any more oil leaking, until I made it to a Philadelphia International truck dealer that night. It was a damned good thing the shop was open until midnight.)

One of NYC's fine policemen arrived about an hour after I called, which I didn't think was too bad, considering all the congestion. He made notes of my version of the accident, and then asked where the station wagon's driver went. I informed him of the driver's parting farewell before he left, and the cop said something like, "Fucking unbelievable." When I handed him the station wagon driver's information, the cop said, "Well, the son of a bitch shouldn't have left!" He then gave me a copy of the accident report and a copy of the citation he was writing for the guy.

The following Monday morning, I was sitting in WMTS's safety director's office, filling out an accident report, when the safety director, Bill, received a phone call, which he quickly put on speaker. The very pissed-off station wagon owner loudly proclaimed, "Your truck driver destroyed the whole left side of the company car my employee was driving last Friday. What are you going to do about it?"

Bill grinned at me, and then calmly replied, "Is your driver's name

Dino Avenelli?"

He said it was.

"Sir, if you will contact the precinct, you will learn that your driver has an outstanding warrant for his arrest for leaving the scene of an accident."

Bill provided the owner with the citation number, and then we could hear the owner giving hell to his employee.

"Sorry to have bothered you," he said curtly before hanging up.

Of course, not all cops were helpful. One I encountered was more chickenshit than anything else. I had entered the Pennsylvania Turnpike at Breezewood, and then headed west toward Pittsburgh. The state police weren't too creative around this location, and every trucker knew that a trooper was usually hiding behind the bridge close to the police barracks, and sure enough, when I hit the designated spot, the cop was where he was where he was supposed to be. Anticipating this, I had made sure I was driving at the speed limit, which is why I was surprised when he pulled out, turned on his blue lights, and pulled me over.

I got out of the truck and walked back to his car—you were allowed to do that then—and he informed me that my trailer marker lights weren't working. He then followed me back to the front of the trailer, where I climbed up onto the catwalk—the steel decking behind the truck cab—and jiggled the pigtail—which hooks the tractor lights to the trailer—and the lights came back to life. I'm not sure why this light connector is called a "pigtail," because it certainly isn't curly, but different wires run from the tractor into this one pigtail, which connects to a trailer and operates its marker, stop, and signal lights. I hopped down and thanked him for letting me know the lights were out, and I turned to head out but he stopped me and curtly told me to bring my license and truck registration to his car. It must have been a slow night, or he must have needed to make a quota, even though lawmen will swear that there is no such thing as a quota, because the bastard issued me a ticket for my lights not working, or some similar infraction. I questioned why he had to give me a ticket, since now the lights were working fine, and he just shrugged and said that I should have jiggled the pigtail before I came around the curve where he was sitting. Due to the fact that I didn't want him to write more tickets, I waited until I had driven away before I called him a hell of a lot of unkind things.

Just about every driver knows there are times, usually an hour or two before dawn, when it is all-but-impossible to keep your eyes open. He also knows how helpful his CB radio can be in staving off his sleepiness because he can use it to shoot the shit with other drivers. Some nights he might talk with another driver for twenty to thirty miles. Other nights, he could travel hundreds of miles with another driver, talking trucks, family, football, or anything at all, really. It helps break the monotony and keep them both awake.

One night I was driving while listening to two drivers chatting away. It was one of those nights with very little CB chatter, and I enjoyed listening to them for quite some time. They were talking casually, without all the foul language we truckers hear all too often, and other drivers were quietly listening too. Except for an occasional interjection with a smokey alert, we mostly just listened. Then, at one point, the driver of the rear vehicle came on the CB and excitedly shouted, in a very high-pitched voice, to the driver of the leading vehicle, "Hey, man, there's something running alongside your trailer! I can't see exactly what it is, but the son of a bitch is running right beside you! Oh, oh, oh, goddamn, wait a minute. It's, it's, it's . . . it's your motherfucking tie-yah!"

One of the lead driver's trailer tires had come off its axle, and the tire's momentum caused it to keep pace with the trailer until it eventually ran off the road and headed into the highway median. When I passed the scene a few minutes later, I could tell that the drivers had retrieved the offending "tie-yah" out of the ditch and were rolling it back across the highway. A few miles down the road, the CB radios suddenly came back to life and for the next hundred miles, drivers were laughing and repeating, "It's your motherfucking tie-yah!" I must have heard it a dozen times, a symphony of "tie-yahs." I may have even repeated it once or twice myself.

When I arrived at my destination in upstate New York, I delivered a load of empty one-gallon glass jars to an apple processor, and couldn't wait to crawl into the bunk because I was plumb worn out. I handed the shipping papers to the receiver and asked if he could wake me when they finished unloading my trailer. The guy looked at me and politely stated, "Well, you have to help us unload your trailer, so we can just wait until you finish your nap before we start on it."

Shit, there wasn't much I could do about it, so I just figured it would be unloaded sooner by my helping, which would get me in the bunk sooner, with a full night sleep instead of a dinky nap. The empty jars were packed four to a case and all we had to do with this floor load was to place each case on a roller conveyor. Unloading actually went quicker than I expected—the unloading crew seemed to work faster because I was working fast and steady. After, the receiver told me the jars would soon be filled with apple cider and asked me to follow him into the warehouse. When we got there he hit a button that opened a thick, heavy door to reveal a cavernous refrigerated warehouse packed full of bins of gorgeous red apples. Each apple had very fine droplets of moisture on it. The fellow told me the apples had been in storage for several months while waiting for processing. He found a bag, filled it with those luscious-looking apples, and handed it to me, and then wished me a good sleep and a safe trip.

Later, I phoned my dispatcher and he gave me my next assignment, telling me to pick up my next load after I woke up. I was to deadhead (drive empty) just west of Pittsburgh to pick up a load of motor oil that needed to go to Atlanta. The shipper was open 24/7, so it didn't matter what time I arrived. The trip from New York to Newell, West Virginia, and then to Atlanta was so uneventful that I probably wouldn't even remember it if that receiver hadn't given me that bag of apples, and especially if I hadn't eaten five of them along the way.

My early delivery in Atlanta went very quickly, and afterward, I started heading north to a truck stop off I-85 to grab a shower and some breakfast. All of a sudden, my stomach started growling so loudly that I heard it over the truck noises. Then, the cramps began. *Oh God, I beseech you, where is that exit for the truck stop?* A sign read five miles to the exit and I breathed a sigh of relief, I was almost there. For a bit the pains began to subside, but one mile later, the cramps returned with a vengeance, and my rear end was ready to explode. *Dear Lord*, I begged, *if you don't let me shit my pants, I promise I'll never eat an apple again.* I clinched my cheeks so hard that my body went stiff as a board, which made driving the truck anything but safe. It was like I'd taken a malaria pill.

Finally, the exit appeared. But as I started up the ramp, I saw a long line of trucks waiting to turn onto the highway leading to the truck stop. The pain was unbelievable, and the wait would be excruciating. What's a trucker to do? Knowing I could not hold it any longer, I pulled onto the ramp shoulder, set the brakes, turned on the four-ways, grabbed a box of Kleenex, jumped over the guardrail, and practically ran down the embankment until I found a somewhat level place to squat. That feeling of instant relief was unlike anything I had ever experienced or will ever experience again. I was so weak that I had a hell of a time climbing back up the hill to my truck. I have often wondered what everyone thought about the fact that I couldn't stop grinning the whole time I showered and ate breakfast. You can find happiness in even the direst times.

I drove through the night to make my delivery in Eastern North Carolina, and arrived empty at the WMTS home office around lunchtime. I couldn't wait to crawl into the bunk to catch up on my sleep, but before I did, I walked into the office to turn in my paperwork. I was informed of an extremely "hot" paper load, meaning it needed to be delivered ASAP. It absolutely, and without fail, had to be at the receiver's printing plant in Jonestown, Pennsylvania, the next morning. The printing plant was out of paper and the shipper had made the urgency very clear. I was told that I would be hauling the load, and in order to get a better rest than I would have gotten in my truck's sleeper bunk, the company checked me into a local motel, gave me a car, and told me they would call me when the shipper finished making the product later that afternoon. Another driver would bring it to the terminal, and then I would drive like hell to get it to Pennsylvania by seven the next morning.

Murphy's Law—that anything that can go wrong, will go wrong—came into play when the relay load did not even arrive at the home office until close to eleven o'clock that night. I had already fueled the truck, so all I had to do was hook up and haul ass, but Murphy must have also hitched a ride in my passenger seat, because I had driven less than sixty miles when I noticed the first snowflake. Each additional mile brought more snowflakes. I was very glad my seventy-seven-thousand-pound vehicle helped my traction as it snowed, and snowed, and snowed. I had driven through snow many times before, but never anything like this, and not for this duration.

Just after daylight, the heavy snowfall finally eased up, and I was slowly proceeding up US 15 N, just south of Gettysburg, Pennsylvania. I was driving in the well-packed-down right lane when I noticed a produce-hauling truck coming up behind me in the left lane. The driver was lugging along through what was now twelve inches of snow, going fifty to fifty-five miles per hour in a late model black Marmon. I was impressed, wondering, *How could he be running this fast, when my lane is as slick as molasses?* Hoping to learn from watching him, I moved several feet left from my lane and into the lane that still had deep snow, and by God, I got educated on the proper way of running in snow: It was necessary to get out of an ice-packed lane and into the snow-covered one because the deep grooves of a tractor's drive wheels get more traction in snow than they do on ice. With traction, I was able to pick up my speed enough that I stayed with him for quite a few miles.

At 7:45 that morning, I was just a couple of miles from the consignee. I sorely needed a cup of coffee, so I pulled up on the right side of the road beside a large mound of plowed snow, and walked into a McDonald's. When I opened the door, everyone in the place, including the employees behind the counter, was staring out the windows. I turned around to see what captured their attention, only to see that it was my rig that had them mesmerized. The tractor and trailer looked like a solid block of ice and snow. The snow and salt, which I had traveled over all night, had been thrown upward and caked underneath the trailer and behind the tractor. All openings had been filled with snow. The rig resembled a big, white locomotive, right out of a Coors Light ad. You could hardly tell the tractor's make or color. It was one of the strangest sights I'd ever seen.

The snow caused me to miss the seven o'clock delivery time by an hour. When I handed the bills of lading to the receiver, I said something like, "Sorry I'm late, but here's that hot load of paper everyone's been screaming about." He looked over the listing of the load's contents and proclaimed, "We don't need this shit. We have a whole warehouse full of it. The damned paper mill didn't even send what we needed."

Every truck driver has had similar shipper snafus happen to him. Upon delivery, we learned that our "hot" loads really weren't hot; or sometimes, when we took our time getting there, the loads turned out to be "hotter than hell!" Most drivers realize that these scenarios happen, and probably will again sometime down the road. Although we wish we'd known the facts before we began the run, we didn't, so there isn't much we could have done about it. About the only thing you can do is pat yourself on the back for doing your part.

The following morning, the weather was much warmer than it had been in Jonestown, Pennsylvania. As I drove my white locomotive across a railroad track in Eastern North Carolina, the uneven roadway broke loose several big chunks of the caked-on snow from under my trailer. The area had received no snow that previous day, and I wondered what drivers thought when they observed those icebergs laying in the roadway before they melted in the warm weather.

One Friday afternoon, I had loaded a full load of wooden pallets from a manufacturer in Warrenton, North Carolina. There were so many stacks of pallets that I had to use eight nylon straps, in addition to several chains and binders, in order to hold the load in place. As with the finished lumber, you couldn't tighten the shit out of the pallets, you could only tighten them snugly.

I was taking the rig to the home office yard for the weekend, so I traveled down US 401 for several miles until I reached NC 58 S, which required a left-hand turn in the very middle of Warrenton. I had a green light, so I pulled into the intersection as far as I could to allow enough room for the trailer to make the left-hand turn. Opposing traffic was still traveling through the light as I caught the eye of an extremely beautiful young woman as she crossed the street. She looked college-aged and had long brown hair. It was summertime, which meant that it was Eastern North Carolina hot. If memory serves, she was wearing white shorts, a tight, Carolina blue T-shirt, and black flip-flops. Other than that, I don't remember much about her. She smiled at me, and, of course, I returned a smile.

As the stoplight was changing to red, a car from the opposing traffic lane passed by my tractor. The young lady had reached the sidewalk by the time I began my turn, and since she was still smiling at me, it was only polite to smile back as I made my turn onto NC 58. All of a sudden, my rig started slowing, so I looked back from the driver's window to notice my trailer wheels trying to climb over the trunk of that last vehicle to go through the light. Traffic had backed up in front of the car, and with nowhere to go, the car was sitting in the middle of the intersection.

I am pretty damned sure I was responsible for tons of cuss words that Friday afternoon, because I had tied up Warrenton's Main Street at quitting time. Thankfully, no one was injured, although my ego certainly took a hit due to the young lady's seemingly uncaring feelings. She lost her smile as she watched the accident happening, and she stayed on the scene long enough to look at me, look at my trailer, look at me again, and then shake her head in disapproval. Then she continued walking down the sidewalk as though she couldn't have cared less. If I could have read her mind, I'm sure I would have heard her thinking, *dumbass.*

Warrenton's weekly rag probably read, "Local beauty queen distracts truck driver, causing accident on Main Street during rush hour."

I wonder how many accidents truckers have had while appreciating the sights around them—especially during warm weather.

For my final college trucking trip, I was dispatched to Indianapolis to pick up a load of pinto beans. At the bean processing plant, the dock foreman asked, "Where's this damned load of pinto beans going? North Carolina?" I was surprised at his foresight and asked how he knew the destination.

He replied, "Cause that's the only goddamned place people eat them fuckin' pinto beans!"

He and his buddies laughed their asses off. *Hell,* I thought, *I do love pinto beans,* so maybe he was right.

This was a floor load and I had to help the workers transfer fifty-pound bags of pinto beans off their pallets and then stack them on the floor of the trailer. The delivery appointment was scheduled for early the next morning, so after sweating my ass off loading the beans, I drove a few miles to a truck stop, where I showered and got something to eat—which definitely was not pinto beans after humping those fifty-pound bags of them.

Twenty miles into the trip south from the truck stop, a Good Samaritan driver called me on the CB to inform me of a flat tire on my trailer. I soon found a truck stop with a tire repair shop, but I was told that I might have to wait several hours for the repair. The hot weather was causing many blown-out tires and the shop was extremely busy. There was nothing I could do but wait, but while waiting I began having doubts as to whether I could stay awake long enough to make my delivery on time the next morning. Over coffee, another driver said he had a couple of "truck driver pills" he could sell me that would help. I wasn't so naive that I didn't know about black beauties, but I'd never used them. I decided, however, to give it a go, so I paid him, and gulped a few down.

Back on the road, I headed south, and it didn't take long to realize the pills were beginning to work. My mouth was as dry as a pinto bean fart, and I was smoking Camel non-filters like they were going out of style. Just before I crossed from Kentucky into Tennessee, a loud "*pow!*" alerted me to yet another blown tire—this one on the tractor. I had to limp several miles before I could pull over at the Tennessee Welcome Center; but, unfortunately, I had used my only spare tire back in Indiana. This was the 1970s, when we had to use pay phones and phone books, and I spent a very long time and shitload of quarters trying to find road service. There was no tire service available at midnight, and the very best any company could do was to bring me a tire at seven the next morning.

This meant that I was shit out of luck, and again there was nothing to do but wait, so I took my shower kit into the men's room, washed my face, and brushed my teeth. I crawled into the bunk and laid my head on my pillow, determined to sleep until the tire man made his appearance. But my eyes wouldn't shut. The pills were doing such a fine job that I couldn't have closed them even if I'd used a pair of heavy steel pliers. At the time, there were no truck televisions, smart phones, or wireless internet access—hell, Al Gore hadn't even invented the internet yet—and I had nothing left to read; I'd already finished the one book I'd taken on this trip. So I tossed and turned, and eventually became so fidgety that I had to get out of that bunk because I was getting the heebie-jeebies.

I plopped myself down at one of the many picnic tables lining the sidewalk leading to the welcome center and struck up a conversation with practically everyone who stopped at the rest area. It was summer and there were many travelling vacationers, so there was no shortage of folks to talk with. I greeted most everyone walking into the welcome center building by asking any number of questions. They certainly didn't need to know that I was unable to sleep, or why, so I simply informed those who asked that my truck was broken and couldn't be fixed until the following morning and I was biding my time. I learned that several families were taking their kids to college, while many others were going to Myrtle Beach, South Carolina, for vacation. Most everyone was quite pleasant and seemed to enjoy a bit of conversation. A few folks, though, were not so agreeable, like the one fellow I asked, "How are you doing this fine evening?"

He gruffly replied, "It's none of your damned business. Leave me alone!"

Several others just stared at me and said nothing when I asked them how they were doing, but, not to be deterred, I kept asking the questions for hours.

One thing I figured out quickly was that about three out of every four men entering the men's room didn't really need to use the facilities. Rather, they were there seeking the company of other men. I wasn't thrilled to be propositioned a couple of times during the night—and one guy didn't want to take no for an answer. He wouldn't get up from the picnic table when I told him I wanted to be left alone, so I got up and began heading back to my truck. Somehow he misread this, as if I was inviting him to come with me, and he began to follow me, but he quickly ran off to his car when I pulled my tire thumper, a miniature baseball bat, from under the seat. When I went back to my perch at the picnic table, I took it with me, and I kept it with me the rest of the night. With it in hand, I didn't receive any more offers—but maybe holding that thumper was also the reason no more vacationers stopped to chat. So I drank Cokes, smoked Camel non-filters, watched hundreds of vehicles come and go, and walked around the picnic grounds by myself until morning. I learned that day that black beauties will do a fine job of keeping you awake, but when it comes to the places where I've seen the sun rise, I can think of a hell of a lot more picturesque settings than the Tennessee Welcome Center.

———

Either life got in the way or I was so broke that I couldn't focus, but for one reason or another, I didn't finish college. At the time, I was driving a piece-of-shit VW camper, which broke down pretty often, so when WMTS offered me a well-paying job at its biggest terminal, in Baltimore, Maryland, I gratefully accepted—and so began my fulltime career in the trucking industry, which I began as an assistant terminal manager.

PART FOUR

CHARACTERS

FRANK

Frank was a WMTS guy I'd met before I went full-time. He had a lot of driving years under his belt and when we met he was WMTS's mechanic, road service person, and forklift operator. Like Obie, he was a very capable jack-of-all-trades. What he lacked in formal education was eclipsed by his intelligent application of common sense. There was no mechanical problem he couldn't solve. He approached each issue, studied it, and then resolved it.

And Frank never shied away from any job that had to be done. Back when I was in college and working for WMTS, I had fueled up at the Baltimore terminal late one very cold and windy winter night and driven up I-95 North about thirty miles, on my way to somewhere in New England. Just as I was passing the Maryland House rest stop in Aberdeen, Maryland, my truck's main radiator hose blew out, and most of the antifreeze was lost. I contacted Frank and he said he would help me as soon as he could get there. Luckily, I had both a blanket and the military poncho liner I had traded for in Vietnam, so I stayed tolerably warm in my bunk if I didn't move around too much.

I had to get out of the bunk when Frank arrived, because those were the days that the entire tractor cab had to be jacked up and tilted over in order to access the motor. I was so damned cold that I sat in Frank's pickup truck most of the time, but Frank braved the cold and masterfully installed a new radiator hose and filled it with antifreeze, even while his hands were frozen stiff. When he was finished, he started my tractor, left it running so it would warm up before I headed north again, and then got into his pickup truck. He sat there and shivered for a long time, never saying a word about how cold it was outside. Watching him was a lesson in perseverance.

Shortly after I began working in the Baltimore terminal, we had an extremely busy day that kept us at work until after eight o'clock that night. When we finished, Frank and I decided to grab a beer and a pizza before heading home in opposite directions. We rendezvoused at a country music bar, which Frank referred to as his "shit-kicker hangout," and several beers loosened our tongues as we talked about our personal lives.

When I learned Frank was on his second marriage, I asked what had caused his divorce. He gave a sigh and a half-smile and said it was because he had contracted the "seenus disease."

"You mean sinus disease?" I asked.

"No, I was in the back seat of my car with this 'ol gal, and my wife SEEN US. I don't know how she found me, way off down a dirt road, but let me tell you, it scared the hell out of me when she banged on the rear window. I got my pants on, stepped out of the car, and asked her what she wanted."

Frank's wife asked him who the woman in the back seat of his car was and he responded that he didn't see a woman in the back seat of his car.

Not long after, at Frank's divorce hearing, he again claimed there hadn't been a woman in the back seat of his car—but the judge didn't buy it. He left the marriage with little more than the pants he'd pulled up when he exited the back seat. I'm sure he's not the only man who's suffered from the seenus disease.

Frank was a master of words and there was no filter, particularly when it came to women. I was riding with him one day when he saw a gorgeous, well-proportioned woman crossing the street at an intersection. He shook his head several times and said, "My, my, my. That gal could put something on you that Ajax wouldn't take off—and that Ajax is good stuff!" He loved this Ajax proclamation, and applied it to anything that he had a strong appreciation for, from well-proportioned women to Gojo hand cleaner, a great sledgehammer, WD-40, and even a bottle of Jack Daniels. But he also had words for women whose bodies weren't perfect. About one extremely, and I do mean extremely, overweight woman, he said, "If someone told her to haul ass, she'd have to make two trips!" About another overweight lady, he said, "If someone told her to haul ass, she'd have to use a wheelbarrow!"

He truly had words for everything. When he heard snow was in the forecast, he'd say, "I hope it snows so damned deep that when you shit, you'll have to put it in a shotgun and shoot it out the chimley [sic]!" Frank's comment sure suggested a unique solution for what to do if the snow was so deep that you couldn't open the door and get to the outhouse. Snow elicits comments from us all, but everyone at the terminal learned to keep the word "snow" out of conversations, otherwise, you knew what you were about to hear. I think I heard the shit-and-shotgun saying at least a thousand times. Other people tried to use it and bombed, and although we tried to avoid prompting it from Frank, he definitely had a knack for making us chuckle every time he said it.

Frank never used old, worn-out clichés—he made up his own. If easy-to-fix problems turned complicated, or if someone changed something midstream, most people would say something like, "Now things are becoming more interesting." Frank would say, "And the cheese gets more binding-er." When folks heard this saying for the first time, they were at first baffled, but after thinking about his words, it didn't take them long to grasp his real meaning. Given an experience I had one night out with Frank, when I took a bite of fresh out of the oven, 450-degree pizza too quickly and it got stuck to the roof of my mouth, I certainly understood, in this incident, that "binding-er cheese" was about as interesting—or dangerous—as it gets.

At the Baltimore terminal, we had a forty thousand-pound capacity Pettibone forklift, which we primarily used for repositioning steel when axle weights needed adjusting. Another use was when one of the steel mills thought its plants would be hit with a work stoppage—for a technical reason or an employee strike—and would ship ahead of time, and ask our company to stack the steel in our warehouse and on our yard until it was ready for delivery. During these times, the forklift was very busy unloading pipe, plates, tinplate, and coils into our storage spaces. One winter, Frank fabricated a workable snowplow that fit into the long blades of the Pettibone, and after that our yard was always plowed clean.

Sometimes moving a heavy single steel coil proved too heavy for the forklift's counterweights, and the rear wheels, which steered the forklift, would come off the ground. This made coil movement pretty tough and when this happened, Frank would search the driver's lounge for four or five of the largest drivers to help. The group of guys acted as a counterweight, and they'd stand on the rear of the forklift, which usually added enough weight to get the job done. I am certain that the Occupational Safety and Health Administration wouldn't have approved of the human counterweight method, but it worked for us. Frank described this action as "putting something on the forklift that Ajax couldn't take off."

All his pithy pronouncements were intended to make people laugh. Even if he dropped a few offensive lines here and there, he wasn't a mean person. He was extremely careful that none of the folks who inspired his comments ever heard the words he uttered. On the outside, he was a tough-sounding, gregarious, happy-go-lucky, fun-loving guy. Inwardly, he was one of the most compassionate men I have ever known.

SLICK JACK

On the opposite end of the human spectrum sat a worthless little weasel named Jack. Jack was our evening dispatcher, and one of his duties was to reimburse company drivers for their road and truck expenses. Each driver was given one hundred dollars at the beginning of their workweek to use for such things as tolls, tire repairs, and motor oil, and could replenish their funds by "cashing in" receipts for the expenses. Jack was employed by WMTS and working at the Baltimore terminal before I was transferred there, and after I settled in it didn't take me long to notice how he operated. He slinked around and always looked like he was scheming to be one step ahead of everyone else. You know the feeling when you meet a person who makes you think that something about them is just not right? I had that feeling often when it came to Jack.

Anyone who has ever been responsible for a petty cash box knows that it's not a fun task, and try as you may, you cannot always get the son of a bitch to balance. Well, Jack's petty cash box always balanced. Maybe this was one of the reasons I started noticing that the amounts of toll reimbursements were higher than usual, especially the first day of each week. This was odd because most of the company drivers began each week with all of their one hundred dollars of expense money, so they would not have had the need to use much of it.

I was responsible for the petty cash duties during the day, prior to Jack coming on duty at four o'clock in the afternoon, and one day a driver came into the office and asked if I could give him change for a twenty-dollar bill. I opened the box, counted out twenty dollars using various denominations, and then handed it to him, noticing a grin on his face. As he stood in front of my desk after handing me the twenty-dollar bill, I started to place it in the box when I noticed something peculiar. Further inspection showed this to be a humorously counterfeited twenty-dollar bill. I can't remember what the image was, whether it was the classic fake dollar bill with *Mad* magazine's Alfred E. Neuman where Andrew Jackson is supposed to be, or something else, but it was certainly something of that nature. Laughing, the driver handed back the change and told me to keep the fake bill, as he had plenty more.

Jack went about his routine when he got to work that afternoon.

He always began his shift by going into the driver's lounge to see which drivers were available for dispatch, and note who was laying over for the night. As he walked out of the office, I had an idea. I folded the twenty and placed it under a bathroom door, leaving visible only one corner of the fake bill. I told the other office folks what I was doing, and we watched "Slick Jack"—as we not-so-affectionately referred to him—smoothly pretend to drop some documents and then pick up and palm the fake bill, none the wiser that we'd all seen what he'd done.

I think the majority of people are honest, and I hope that most folks, when faced with a similar situation, might say, "Damn! Look what I just found!" or "Wonder who dropped this twenty-dollar bill?" Not Jack. He pocketed it without a word. I wish I could have seen his face when he got a good look at it. All the better if he got caught trying to spend it.

Close to a week later, this petty cash thing was bothering the hell out of me, so I began comparing toll ticket time stamps with tractor numbers. I realized that each driver was required to write his tractor's number on the top of all his receipts and, eureka! That solved it! Jack was a cheater, and I had proof. The time and place of some of the ticket claims did not correspond to the indicated trucks. For instance, one toll ticket showed a certain tractor going through one of Richmond's toll facilities early one morning, but when I went through company dispatch records, they showed that the tractor had been making a delivery south of Atlanta that day.

When other office employees and I investigated, it didn't take us long to figure out how Jack was sneaking cash. He had arrangements with several of our owner-operators and would buy the O/Os' receipts for half the face value of each. WMTS's policy was to only reimburse employees' expenses, not those of O/Os, so they would normally have had to pay the full cost of each themselves. But Jack's scheme was to buy the O/Os' receipts for half the face value of each, so they made half their money back. Then he reimbursed himself from the petty cash box for the full receipt amount. Today, some trucking companies reimburse their O/Os for toll fees, but this was not the case in the 1970s.

When Jack arrived for work the afternoon after we figured out how he was playing the system, I told him to sit down. He said to let him first get a cup of coffee, but I informed him that there was no sense in pouring a cupful since he was not going to have time to drink it. I don't remember my exact words, but they had something to do with him being a lying, stealing, sneaking piece of shit. He began protesting and swearing his innocence, but I told him to shut up and keep quiet until I finished what I had to say. I made him aware that he was fired and would soon be arrested and charged with theft of company petty cash funds. Jack said it was bullshit and began babbling about suing WMTS for unlawful termination, blah, blah, blah. We didn't want the scumbag to receive unemployment benefits upon his termination, so I informed him that the company would act graciously by allowing him to voluntarily resign, although this offer was only good if he accepted the offer immediately by hand-writing his resignation letter. Jack yelled that he couldn't be railroaded into quitting, but changed his mind and decided to write the letter after I picked up the phone and told him I was calling 911.

When Slick Jack wrote his resignation letter, he was probably thinking he would receive unemployment compensation due to the fact that Maryland, where we lived and worked, had the reputation of awarding compensation to damned nearly everyone who applied for it, without much inquiry into the reason for termination. And true to form, he proved us right. We learned that Jack left the terminal and marched right into the local unemployment office. We were, however, delightfully surprised by the fact-finding phone call from the office several weeks later. Jack told the unemployment folks that he had been fired without cause. We supported our version of the story by faxing the unemployment compensation office a copy of Jack's signed resignation letter.

There is a happy ending to the Slick, Sneaky, Stealing Jack story because the unemployment office denied his request for benefits. We never heard another word from the son of a bitch.

EDWARD

Every morning, Joe, the terminal manager, and I would go out onto our lot to wake the drivers. A food wagon came by each morning, and most of the drivers wanted to be woken up so they could get breakfast. One day Joe knocked several times on the driver's side door of Edward's truck, but Edward didn't respond. Finally, he pounded loudly, and hollered, "Goddamn, Edward, if you can't hear me, you must be dead!"

After waking the last driver on the back of the lot, I headed back up front and saw Joe frantically waving at me. He was gesturing wildly and pointing to the truck, but couldn't get a word out even when I reached him, so I climbed up into the tractor to see the problem. Behind the privacy curtain, I found Edward. He was dead. His autopsy revealed that he had died in his sleep from a heart attack.

Poor Joe was terribly ashamed for saying Edward must have been dead if he couldn't hear him. It took him quite a while to get over Edward's passing.

The night before his death, Edward had joined several drivers in our terminal's driver lounge for a low-stakes poker game. These frequent gatherings provided entertainment for the fellows who didn't want to go out on the town. The drivers who had played cards with Edward said that he had kept his usual $150 to $200 in his pocket, so, hoping to give this money to Edward's family, I contacted the coroner's office to inquire when I could retrieve his belongings. According to the two coroner's office employees, the ones who removed Edward's body from his truck, the driver's wallet had various items in it, but no cash. This seemed mighty fishy to me, begging the question of whether the coroner's office had its own Slick Jack.

I never received a reply to the nasty letter I sent the coroner's office, and I never learned the names of the two worthless scoundrels who picked up Edward's body. I suppose it was Edward's bad luck to have died within the Baltimore City limits.

DAVONNE

A week after Edward's death, we asked one of the guys to drive Edward's truck back to our home office in Eastern North Carolina, some three hundred miles away. Davonne said, "Oh, hell no! Edward done died in that truck and his ghost might still be in there!" It took an awful lot of negotiating, including what could have been best described as hazardous duty pay, before he finally agreed to the assignment. As he left the terminal, his parting words were, "Now I'm gonna do this for y'all, but I ain't stopping for nothing after I leave here!"

He kept his word, and I'm sure Davonne must have broken every speed limit along the route, because he made the trip in record time. We later heard that after dropping the truck at the home office, Davonne literally jumped out and ran away from it. He later told someone he had pulled the privacy curtain closed and didn't once look into the bunk.

JB AND WILLIAM

Many of WMTS's drivers from its Eastern North Carolina headquarters spent at least one night each week in Baltimore, while waiting to load early the next morning. It sometimes happened that a few drivers would have to layover twice in one week. When they laid over, most drivers played cards or watched television, but some drivers enjoyed Baltimore's nightlife.

JB was a driver who didn't mind staying in Baltimore, which was mostly due to his fondness for a couple of ladies at one of the clubs close to the terminal. Truth be known, JB probably planned out his schedule to include twice-per-week layovers. Although I never observed him in his element, I heard that he had a way with the ladies. Evidently, they loved his muscular build at 6′3″ and 220 pounds. And I know he was a smooth talker, since he always tried to sweet-talk the dispatchers into giving him better loads than what he'd been assigned. After he spent a night out, it wasn't uncommon to see JB brought to work early the next morning in a car driven by one of his lady friends.

JB was accompanied to the club one evening by a newly hired driver named William. William was young, and the complete opposite of JB. He clocked in at around 5′5″ and 125 pounds, soaking wet, and was so naive that it seemed to drip from him. Poor William also had a speech impediment you heard each time he began a sentence. It was something between *tsk-tsk* and *t-t-t*. I'll just call it *tk-tk-tk-tk-tk-tk*.

William evidently liked what he found during his first visit to the club with JB, because he couldn't wait to return the following week. The morning after his second visit, he had a big grin on his face when he came into the office to get his pick-up slip for his load at the steel mill. I don't know why none of us asked about his "big night" the evening prior—maybe we were too polite. At any rate, William picked up his load of steel and headed south without sharing.

Well, one of the company officers was known to enjoy the occasional practical joke and the next afternoon, the officer told the home office dispatchers that he needed to speak with William when he unloaded his steel and made his "empty" call. Those were the days before cell phones or in-cab communication, which meant every driver, after unloading, would either use a consignee's telephone or have to seek out a pay telephone, in order to inform their dispatcher that their truck was empty because they'd made their delivery, and were ready to receive their next loading assignment. When William called later that day, his call was transferred to the company officer who acted very serious on the phone, saying: "William, some woman called here around lunch today, and she said that she would have called you directly, but she didn't know how to get in touch with you. She said she was really sorry, but she had gone to the doctor that morning because she hadn't been feeling very good. The doctor told her she has a fatal disease and that she doesn't have but about six months to live. William, she said she was afraid that she had passed it on to you. She said there isn't a cure at this time, but she wanted you to know about it because you are such a nice boy. She said she wanted you to know while you still had some time left."

When the officer finished telling William about his fatal disease, he said that William did not miss a beat as he replied, *"tk-tk-tk-tk-tk-tk*, well, sir, *tk-tk-tk-tk-tk-tk*, JB's gonna die too!"

RICKY

All trucking company employees, whether drivers or office personnel, have always been known as hardworking men and women. Their reputation also has them known as hard drinkers when they're off the clock.

One of WMTS's drivers, Ricky, certainly enjoyed his alcohol, but never while he drove. When sober, he was a driver you could usually count on to deliver and pick up on time, but every several months, Ricky would tie one on and go AWOL. Each time this happened, we knew exactly where to find him, so we would put two drivers in a truck headed south from Baltimore, and one of them would drive Ricky's rig back to Baltimore. The drivers would find Ricky's tractor trailer sitting in a parking lot on Highway 301, just south of La Plata, Maryland, and Ricky would be across the highway in a motel, better known as his "drunken home away from home." Sometimes the motel owner would call our office to let us know Ricky was a guest, and that Ricky had given the owner the keys to his truck. The owner always assured us the rig would be safe. The guys would pick up the truck, leaving Ricky at the motel, and several days later, after achieving sobriety, he would call for someone to come get him. We would ask a northbound driver to pick him up and bring him to Baltimore to get the truck. After each "episode," Ricky would stay sober for a couple of months before the bottle lured him back to his motel room.

Through the years, I have often thought that Ricky's drinking habits would have fit in well with those of Obie's crew. Ricky was a nice guy and I hope things worked out for him—that he didn't do anything stupid while drinking.

JACKSON

I'm not certain why bad things happen to certain people, whether they're unlucky, stupid, or just use bad judgment, and other people who do stupid things are unscathed. One guy, Jackson, was an elderly driver by the time we met, and he'd done pretty much everything you can think of. He had always been there, done that, or been there, done that twice, and he had plenty of T-shirts to prove it. No matter how many miles other truckers had driven or how many places they'd traveled to, if you asked Jackson, he had driven more and seen more than anyone else ever could. He was highly opinionated—to the point that Obie would have said, "Jackson would have argued with a stop sign!"

One of Jackson's legs was quite a bit shorter than the other, which made his gait more up and down than forward and earned him the nickname "Hop Along." I remember him saying, "They better not try putting me in a flatbed because of this leg," although no dispatcher would have even considered doing so because he physically couldn't have climbed on and off a flatbed to load, chain, and tarp a load.

One Sunday morning, Jackson arrived at the terminal, ready to begin his trip to New York City. His trailer contained forty-three thousand pounds of paper stock consigned to a printer, and as he and the other drivers filled their trucks with fuel, he shot the breeze with several of them. I am certain the audience allowed Jackson to "hold court." When the other drivers finally said they were leaving the yard, Jackson told them to wait long enough for him to hook his trailer, and then he would run with them up I-95.

When Jackson arrived at his destination in Manhattan the next morning, he climbed out of the cab to open the trailer doors before he backed up to the dock. Well, Jackson opened the doors only to find the trailer empty. His first declarations were that someone was trying to fuck with him, although he later said that the trailer number that he hooked his tractor to was "awfully close" to the trailer number that he should have hooked it to, the one containing the paper, which seems to me to show who's responsible for the missing load.

It turned out that Jackson had pulled an empty trailer for five hundred miles, rather than one with a load of paper weighing forty-three thousand pounds. He had been running his mouth so much, and been in such a hurry to catch the other drivers, that he had failed to open the trailer doors to look inside before he got on the road. He later said that he thought the truck was pulling really well, and figured this was because the maintenance shop had done a great job on it following his repair request. Jackson kept to himself for a few days after the episode, due to all the ribbing he received. He never questioned why he didn't get paid for those five hundred miles.

Another of Jackson's boneheaded moves came about after he picked up a load of beer at a Baltimore brewery. There were no weigh scales at the brewery, so Jackson brought the load to our terminal yard, and upon weighing his rig, he found that the trailer axles weighed in at more than thirty-four thousand pounds, rendering the trailer illegal to drive. Rather than unloading the entire load of beer, and then reloading the pallets in a different configuration to redistribute the weight, he utilized the trailer's sliding tandem.

Most closed vans are equipped with sliding tandems. A sliding tandem is used to move tandems, which are close-coupled pairs of axles, forward and backward to redistribute weight.[1] Depending on the distance between each hole, four inches or six inches, between 250 to four hundred pounds per hole is moved. In Jackson's case, moving four thousand pounds required him to move the sliding tandem ten holes.

When Jackson prepared to redistribute the weight, he began by driving his rig into the middle of our paved lot, where he climbed from the cab and opened the trailer doors, although, when asked later, he was never able to offer any good reasons for opening them. He then released the locking pins, reentered his cab, locked the trailer brakes, and pulled forward, but the tandems did not slide. They were held in place by a scourge of rust. To break the tandems loose, Jackson released his trailer brakes and backed up his rig. When he decided he had attained the proper speed, he locked his trailer's brakes, which accomplished his goal of unsticking the tandems, but the stop was so abrupt that two full pallets of glass beer bottles toppled over and spilled out of the opened trailer doors. They hit the asphalt and almost all the ones that fell shattered all to hell. (We did salvage a few bottles, and took them into the office to keep them safe, but they slowly disappeared.)

---

1 The act of sliding a tandem is uncomplicated enough that a driver can typically perform this maneuver by himself. All that's necessary is to pull the spring-loaded locking pins to release them, and then get in the trailer, lock the trailer's brakes to keep them from moving, and drive the rig forward or backward a few feet to move the tandem whatever number of holes is necessary to transfer the weight. If a trailer's tandems hadn't been slid in a long time, rust buildup made this more difficult, but still doable.

Jackson's excuse for the incident was that it was entirely due to the "damned shoddy shrink-wrapping job" on the pallets of beer, and he declared, "I'll give those bastards hell the next time I go there to pick up another load of beer."

Jackson probably should have been fired many times, especially after this incident of damaged beer. In another industry, his incompetence would certainly have led to his dismissal. The truth is that trucking companies tolerated drivers like Jackson simply because of the shortage of truck drivers. As long as drivers didn't damage their tractors and trailers, or crash into four-wheelers, or knock down low-hanging gas pump awnings, or tear the hood off another tractor while backing into a slot at a truck stop, or blow several tires when they turned too sharply, thereby dragging their trailer through a concrete ditch, then the carriers put up with the occasional product damage by even the most boneheaded drivers.

THE BM OWNER

During one of those young, stupid, fly-off-the-handle situations, I had a disagreement with one of my bosses at WMTS. Rather than acquiesce, I resigned and went to work for another trucking company, as a manager at a terminal in Louisiana, acting in accordance with the old adage that the grass is always greener on the other side. Well, it turns out the grass was hardly green at all.

One time, the owner of the other carrier, which I'll call Big Mistake Trucking (BM), flew in on his private jet to attend a drivers' safety meeting we had scheduled for Saturday morning at our terminal in Louisiana. I picked him up at the local airport, and drove us to the terminal. When we got there, he told me to drive through the yard so he could look at the equipment. I had anticipated his wanting to do this, so the previous evening I'd parked the thirty tractors and trailers into a very presentable arrangement.

When we got out of the car after parking in the fuel lane, the owner walked up so close to me that I could smell his stale breath. It was clear he hadn't brushed his teeth since at least the day before.

He looked me in the eye and said, "What's all those empty tractors doing out there, boy?"

Well, we did have four or five available trucks, so I replied, "Well, sir, we are doing our best to find qualified drivers to put in them."

He thought a moment and then very slowly announced, "I'd rather have a warm body in those trucks than see them sitting on the fence. Do you catch my drift?"

After assuring him that I certainly did, we went into the building for the safety meeting, where the old man promptly pissed off all the drivers. He informed them that he was ending BM's safety bonus program because he felt he was paying the drivers extra money to do something they were supposed to do already—drive safely.

As soon as he said his piece, he turned to me, told me to take him to the airport, and then walked out of the building and toward the car. Most drivers sat wide-eyed, stunned by what they were just told, but one fellow jumped up and ran after the owner. He got very close to the old man and yelled that he had never worked for a sorrier son of a bitch in his life. I stepped in and pulled the driver away from the owner. On the way to the airport, I was instructed to terminate the irate driver's employment within the hour. The driver was waiting for me when I got back to the office, and after learning of his firing, he said he figured it would happen, and that it really was okay, because he didn't want to work for that mother-fucker anyway.

Well, we got busy hiring drivers, and it only took us about two weeks to fill the empty tractors with warm bodies. Most of us were not at all surprised that it took less than thirty days before three of the five tractors were wrecked and torn all to pieces by those warm bodies. And why the hell not? They wouldn't have received a safety bonus even if they hadn't wrecked their tractors.

It was pretty ironic that the owner wanted more truckers because it became clear very quickly that he enjoyed firing people, and, like with his firing of the driver after the safety meeting, didn't give it a second thought. One time, he passed one of his trucks as it was traveling down an interstate highway. Trucks have quite a bit of room in them, so the driver of this rig, in his effort to be comfortable, was driving with his left foot resting on the dash of the truck. The owner clearly didn't approve of this and after somehow getting the driver to pull over onto the shoulder of the highway, he fired him on the spot, just because of the way he'd been sitting.

Another time, he was in his office at corporate headquarters when he saw something that pissed him off without good reason. He'd designed the office so it had one-way glass, similar to what police interrogation rooms have, and from it he watched a driver pull into one of two fuel lanes he'd installed alongside the office. The driver then got out, put both fuel hoses in his tanks and set them on automatic shut off, and then walked around his rig, thumping the tires to check that none of them were flat. When the old man saw the driver miss thumping several inside tires, he walked out of his office and fired the driver on the spot.

Several years prior to my employment with BM, something happened that I wish I'd known about before I took the job. One Friday evening, BM sent their company planes to pick up all terminal managers, along with their wives, and then flew them to a management meeting at the corporate office. When the husbands and wives finished eating breakfast on Saturday morning, BM gave some cash to each wife and provided them with transportation to go shopping and have a fun morning. The wives were asked to return in time for everyone to have lunch together.

The managers' meeting rolled along smoothly until the owner asked the men if anyone had any gripes or complaints about the way things were done at BM. He said they could be candid and ask about or discuss any topic. Reportedly, one terminal manager did have a complaint, and then another manager chimed in with a separate issue to talk about. Some managers, who didn't have complaints of their own, took sides with other managers' issues.

When the wives returned for lunch, they found their husbands without jobs, as the old man had fired every damned one of them. The lunch buffet went untouched because everyone lost their appetite. Everyone was then taken to their motel to check out, and the company planes flew them back home.

JAMES

A story of the load from hell comes from James, one of BM's drivers who had been with the company for several years before I met him. James was probably fifty years old when I knew him, and he struck me as a great guy. I never saw him in a foul mood, and he had an infectious grin that could make any of us smile. He was also a consummate professional, and always looked like he had just showered, shaved, and put on clean clothes.

The story relates to a load of sheetrock James delivered to Dallas. We'd dispatched him on a backhaul load from Houston to Louisiana, and I'd asked James if he would have a problem hauling coiled steel wire rods using only nylon straps. All of BM's flatbeds were equipped with only nylon straps for securement, and chains were not permitted to be used on the dedicated customer's outbound products. He'd replied that it shouldn't be a problem, because the load only had nine rolls of wire and didn't need to be covered with a tarp.

Any truck driver who's hauled coiled wire rods know that the rolls can move to lay sideways, and poor James learned that nylon straps aren't very effective at securing the product. It takes experience to be able to secure them well with only nylon straps. He had traveled less than one hundred miles before several bundles had already fallen over. When he eased into a truck stop, he found a fellow flatbed trucker willing to help him. Using this driver's chains, they were able to stand the bundles that had fallen back up. Over the next twenty-four hours, James was forced to stop, find other drivers with chains, and then repeat the same process two more times. Thanks to his determination, he safely delivered the load in New Orleans the next afternoon.

When James turned in his paperwork at our terminal the following morning, I added eight hours of on-the-clock time to his road pay before I sent it all to the home office. This was the least I could do to thank him for a job well done, and because the time he spent dealing with the wire rods had caused him to lose the pay of hauling another load that week. I was sure this "attaboy" was the right way for a terminal manager to take care of his driver.

Unfortunately, the adage "The apple doesn't fall far from the tree" became quite evident when one of the sons of BM's owner called me the following week. He said he was looking at the hours I had added to James's pay, and asked if I had checked James's logbook to make sure he was due the extra money. I replied that I had not looked at his logbook, but felt that he was due the extra money because he had done such an outstanding job of getting the load there, in addition to not incurring any wrecker fees.

Thirty-five years later, I still remember the son's reply with crystal clarity: "Check his logs. If he didn't include how he spent the extra time on the wire rods, or if he falsified his logs, then fuck him, and do not pay him!"

Now, there were several reasons I made damned sure James got paid for those eight hours. Most importantly, he had earned it. Secondly, I thought that BM's owner, and both of his sons, were crazy as hell.

Early one morning, two weeks after I paid James the extra eight hours, BM's VP of operations walked unannounced into our office, along with my Mobile, Alabama-based division manager. While I was on a lengthy phone call, I observed the VP pacing back and forth as he chewed on his pipe. When I finished the call, he sat down opposite me and I said to him, "Damn, man. What in the world is wrong with you?"

He sighed, and then said, "I've been sent down here to relieve you of duty." Although I knew exactly why this was happening, I asked anyway. He said the reason for my firing was that I didn't get along well with the management team.

"Well, no shit!" I responded. That much was certainly true.

Just then, the phone rang, and as I instinctively reached to answer it. I stopped myself. I looked at the VP, and said, "You probably should answer it since I don't work here anymore."

The VP had flown in from Mobile and been driven the one hundred miles from the airport to our terminal by my division manager. The VP stayed at the office answering phones while the division manager drove me to my apartment, due to the fact that BM wouldn't let me take their company car home with me. Just two weeks prior, this same VP had suggested that I sell my personal car because I had a very nice company car to use. I'm glad as hell I didn't take his advice.

I had become very good friends with my division manager during my six months with BM, and during our short ride to my apartment, he told me that he had been ordered, three days prior, to travel to our terminal to fire me himself. This fine fellow said he told BM that he wouldn't fire me. He said he hadn't hired me, I was moving all loads on time, my terminal's empty miles were lower than any other terminal's, and I was doing an excellent job. The guy was upstanding and earnest and I believed him. Not surprisingly, BM's management team was not overjoyed by his refusal to follow their orders. They stripped him of his division managerial role, and demoted him to exclusively being the Mobile terminal manager. He quit working for BM a few months later, and I'm glad he did, because we got to work together for several more years at another company.

There was a happy ending to getting canned by Big Mistake Trucking, because two days later, I interviewed for another terminal management position with a flatbed carrier based in Georgia. After leaving this carrier's offices, I found a pay phone and called the president of WMTS to inform him that someone would be calling him for a reference and that he should lie to them and tell them something good!

As luck would have it, he asked that I come to his office the next day to patch up differences with the fellow who'd caused me to leave in the first place. Sure enough, in January 1980, I left Slidell, Louisiana, and went back to work for WMTS; this time, at its brand-new terminal outside Pittsburgh. I was happy to correct a bad mistake. Call it what you may, luck or fate, but trucking allowed me to move to Pittsburgh, where I met and married my lovely wife of thirty-eight years.

PART FIVE

MANAGEMENT

The new WMTS terminal was about five miles from downtown Pittsburgh. The location was perfect in that it was across the street from our largest flatbed shipper—the steel producer that had supported our application for the "operating authority" to carry steel, which, like any other commodity carrying, had to have a shipper as a supporter and be approved by Interstate Commerce Commission.[2] It was also within thirty miles of many of the largest steel shippers in the country, and not far from one of our largest van shippers. At the time, Pittsburgh was so rich in steel shippers that it earned its nickname Steel City. Today, there are hardly any left.

At first, I was a one-man show since I was the only employee at the terminal. I made sales calls, answered phones, accepted freight from steel shippers, and dispatched trucks to pick up loads of steel

---

2   This process, by the way, was bullshit, and terribly expensive. It severely restricted competition, but most trucking companies were happy with the status quo, especially the large LTL (less than truckload) carriers, because rates were high and most companies' profits were good. Eventually Congress realized the rules were arcane and passed the Motor Carrier Act of 1980, more commonly referred to as "deregulation." The large trucking companies hated it because before deregulation, they'd enjoyed a monopolistic hold over shippers and could charge higher rates. Shippers, on the other hand, loved deregulation because freight rates practically dropped overnight. Most everyone else liked the fact that it was now possible for small trucking companies to get their feet through a shipper's front door.

for delivery in southern states. Our drivers also transported lumber and other building supplies to the Pittsburgh area, and typically made their deliveries when receivers opened in the mornings.

I was unmarried at the beginning of my time in Pittsburgh (I met my lovely wife there later), and enjoyed driving tractor trailers when it happened that I did not have any drivers, and these situations invariably took place during nighttime hours.

Many of the late pick-ups were at a steel pipe mill in Aliquippa, Pennsylvania, close to thirty miles northwest of Pittsburgh. After signing in at the gate, I would arrive at the shipping dock on time, although, just like all the other trucks, I had to wait a considerable amount of time before being directed to either pull into, or back up into, a specific loading bay.

In the 1980s, the days of Big Steel, there were fewer restrictions on steelworkers, which meant that loading one truckload of pipes could take anywhere from a half hour if you carried a six-pack of beer to the guys loading your truck and they worked swiftly in return, or four or five hours if you pissed them off or rushed them. Sometimes it took a long time to load a truck because a crew member was on his "sleep shift." Tenured carriers explained to me that after everyone clocked in for their shift, they alternated spending the entire eight-hour shift sleeping in a makeshift bedroom they had fabricated in an empty, secluded section of the pipe mill.

I certainly dealt with my fair share of characters at the WMTS Pittsburgh terminal too. One of our owner-operators, Ed, was parked on the yard one afternoon, and was diligently washing and waxing his truck, while waiting to pick up his next load that evening. When he had pulled into the yard earlier, I'd noticed that his tarpaulins were all balled up, rather than being rolled properly into neat bundles. I surmised that he had unloaded at some location that hadn't had enough room for him to spread out the tarps, thereby making it difficult to properly fold them.

While I spent the afternoon answering phones and dispatching other drivers, I could observe Ed through the office window as he cleaned his truck. I kept wondering when, or if, he was going to fold his tarps, or if he was going to arrive at his next shipper with his tarps still in mess. Eventually, when he didn't seem to be ready to do anything about the tarps, I walked down the lot to where he was and politely asked, "Ed. Are you going to leave your tarps that way when you go to the steel mill later?"

Ed looked at me, grinned and offered, "Damn, man. I was gonna do just that. You must have ESPN!" Maybe he truly did possess ESP, since the network came into existence years later.

One of the most difficult roles of a terminal manager is to be a disciplinarian, which requires giving the ever-unpopular reprimand. Most drivers are aware, unless they're missing those pallets, of why they're getting their ass chewed out, and, truthfully, most ass chewings go pretty well. The times that don't stick with you.

One time, I had to speak with a driver who'd been very rude to our largest shipper. Several other drivers had observed the verbal altercation, and all had reported that the driver had been out of line. The driver's attitude was that the shipper was an asshole, and there-fore, it was his right to get in the guy's face. It was a poor defense and the behavior was inexcusable. I told him I'd have to give him three days off without pay, and if this behavior happened again, we would have to let him go.

At this, the driver jumped out of his chair, stood up, and shouted, "I ought to just whip your goddamned ass and get it over with!"

Now, this big, strapping young fellow could have wiped the floor with my ass, seeing that he had about five inches and over one hundred pounds on me, so I replied, "Well, you can probably whip my ass, but when I finally get up, I will still have a job. You, on the other hand, will be in jail, and you will have no job."

He pondered this for a few seconds, then said, "Hell, you ain't worth it!" and slammed the office door on his way out. He ended up taking more than three days off because, thankfully, he found a job driving for another carrier. I'm fairly certain he didn't join a first-class, upstanding trucking company with strict hiring guidelines, since we didn't even get a past employer confirmation request from his new company.

Part of his bad attitude may have had something to do with his tall, gorgeous wife, since I'm pretty sure he thought she was running around on him while he was on the road. Several weeks before I met with him, his wife had come to the office one Friday during the summer to pick up his paycheck. During our brief conversation, she sat in the chair beside my desk, wearing a damned nearly falling off halter top and marvelously flimsy short shorts. She kept crossing, then uncrossing, her legs, and leaning very close to me, which showcased her ample cleavage. If I was a betting man, I would wager all I have that she knew what she had. She certainly knew how to flaunt it. I have always been proud of myself for not succumbing to the unspoken—yet clearly implied—offer from this fine, redneck hussy—but the memory lingers on, and is still quite vivid. What can I say? We truckers love hot weather.

I left WMTS in the early 1980s and went to work as a terminal manager at another trucking company in Pittsburgh, then worked as the president of a transportation brokerage company for ten years, working in Carnegie, Pennsylvania, and Jamestown, North Carolina, and then went back to the first company to work as a terminal manager at their Baltimore terminal.

While at the second trucking company in Pittsburgh, I met an owner operator named Ralph who one day in May informed me that he was parking his rig because he didn't drive during the summer. When I inquired of his reason for not working, he told me that he didn't drive his truck from Memorial Day through Labor Day each year, when "the terrorists are out on the highways."

"Terrorists?" I asked.

His response was very matter-of-fact, and with a straight, serious expression he told me: "Some people call them tourists, but as far as I can tell, they are out to scare the hell out of me, so I consider every damned one of them as a terrorist." You can't make this stuff up, and, sure enough, Ralph went back to hauling steel after Labor Day.

Another memorable oddball, a far less likable fellow, was a dispatcher I met in Baltimore in the mid-1990s. I'd been at the job a year when our largest steel shipper told me that the dispatcher was not very effective, and wouldn't accept any dispatch responsibility while I was away from the office. When I counseled the guy, I reminded him that he was in charge of dispatch responsibilities when I was away. I was more than surprised when he said, "I wasn't aware of that!" He had evidently forgotten that he'd been performing, or was supposed to have been performing, these duties for the past year that I'd been there, and also for at least a year before I arrived.

I chalked up his odd response to his manner of getting me off his ass, and might have been willing to let things go, assuming he changed his behavior, but then something too abhorrent happened. Upon returning to the office one afternoon, instead of walking in through the front door after parking in front of the building, I dropped my company car off at the mechanic's bay and then walked through the warehouse and entered the driver's lounge. Several drivers were hanging out there, awaiting their dispatches, and I bantered with them some and listened to a few of their jokes. After some time had gone by, I went through a door leading to the office hallway.

As I stepped into the hallway, I heard the dispatcher say, "No, baby. I can't help you with that because that ain't my de-PART-ment! . . . Okay, I will have the terminal manager call you."

I thought to myself, *Oh Lord, please don't let this have been a conversation with our biggest shipper!* When I confronted him, I learned that it had indeed been our biggest shipper, so I placed the call to apologize profusely and had my ass chewed out again.

When the evening dispatcher arrived for his shift, I asked the offending dispatcher to follow me out to the yard, so we could talk in private. I began by telling him how disappointed I was by his actions, and that it was totally unacceptable to refer to someone from the shipper's office as "baby."

He got a terribly wounded look on his face, as though I'd slapped him, and asked, "What are you coming down on me for?"

I let him know I was coming down on him because if I couldn't get him to do the job correctly, the shipper wanted me to find someone else to work with them. I said many other heated things too, and ended it with, "This is the company's largest terminal, and we are moving one hell of a lot of freight. When I am away from the office, if you don't look good, then I don't look good. And let me tell you, I do like looking good!"

With that, he hunkered down his shoulders, cocked his head to look up at me, and declared, "I see where you're coming from! I'm gonna make us all look good!"

For a while after, he acted like a true dispatching professional, and it made me feel great that our disciplining session had elicited positive results; I even started to feel very comfortable with my managerial acumen. Unfortunately, his professionalism only lasted for about a month. He then got so ornery and belligerent that I ended up firing him.

After I informed him of his termination, he claimed that the reason I'd fired him was because he was black and I didn't want a black man to have a dispatch position. I replied that his being black had nothing to do with it, and that I would have fired one of my brothers if he had been as damned sorry a dispatcher as he had been. Our predominantly black driver workforce didn't agree with the dispatcher's comments, and most of the drivers told me that that man should have been fired a lot sooner. The guy was certainly a piece of work.

When I worked at the company's Baltimore terminal, there was a driver who tested positive for illegal drug use. These tests were standard by then, and given randomly. After a driver gave his urine sample at a testing facility, he could go back to driving while the sample was sent to a lab and analyzed. If a driver tested positive, the testing lab's medical review officer (MRO) would contact the driver to inquire why the test was positive. The MRO would talk with the driver before contacting the driver's company.

Apparently, however, the MRO tried to contact the driver who'd tested positive and couldn't reach him. The MRO had left messages on the driver's home phone, and then asked the trucking company to have the driver call him, but the driver did not return the calls. Finally, when the MRO hadn't been able to reach the driver after an extended period of time, the MRO contacted our safety director, Bill, and told him he urgently needed to speak with the driver. Of course, Bill probably knew the MRO needed to speak with the driver because his drug test was positive.

The driver's luck in being able to avoid the MRO ran out a day later when Bill observed the driver getting fuel at a small fuel stop north of Philadelphia. Bill corralled the driver there, and walked him over to a phone booth. He then dialed the MRO's phone number, handed the handset to the driver, and closed the phone booth door. He then watched the driver get very animated during this call, rapidly gesticulating with his free hand while he was talking. After some time, he opened the door, handed the phone to Bill, and walked back to his truck. Bill could hear the doctor laughing even before the receiver reached his ear. When he asked the MRO what was so funny, the MRO said he'd asked the driver if he could tell him why he'd tested positive for cocaine, and the driver had explained that he'd gone to the dentist the day before his drug test. He said the dentist had to do quite a bit a work on his teeth, and he'd been given a "more than usual" amount of lidocaine. He told the doctor that the lidocaine must still have been in his body at the time of the test, and that this is what caused him to test positive, saying "You know, doc, lidocaine, Novocaine, cocaine—they are all in the 'Caine' family!"

Needless to say, the driver was fired, and I have laughed about his excuse for many years.

During my transportation brokerage days, I had one customer, located on the south side of Pittsburgh, who received truckloads of cut steel sheets from a shipper in Kalamazoo, Michigan. I had found only one carrier capable of handling the majority of the three or four shipments per day, due to the fact that the carrier had at least that many trucks per day delivering very close to Kalamazoo. After the first load, the carrier knew the pick up and delivery information, so all I had to do was call each morning and tell them the number of trucks needed that day, and the carrier would dispatch its drivers.

The shipper called me one winter morning with the greeting, "What is this horse shit?" Not sure what horse shit he was referring to, I said, "And good morning to you, too!"

He was clearly pissed off, and told me he'd nearly had a heart attack when one of the "goddamn flatbeds came here to load, and had horse shit frozen to the fucking bed of the trailer."

This was certainly possible because the flatbeds were also used to haul horse manure from horse farms in Kentucky to a mushroom farm near Kalamazoo. The product was baled and tarped for this, but I learned that the driver had driven through freezing weather and substantial rains while travelling through Kentucky and Ohio, and that could have caused problems. It turned out that water had seeped in under the tarps and frozen some of the manure onto the flatbed's floor. Even though the driver unloaded at the mushroom farm, quite a bit of horse shit had stuck to the aluminum floor of his trailer.

The numbskull driver said that he had not even thought about the fact that the shipper might not want its product loaded on top of his trailer-bed full of horse shit. He reminded me of one of those fellows who might be a bit of a numbskull, or, as we described them, "missing a few pallets." Eventually, we located a place in Kalamazoo that had a hot-water pressure washer, and the trailer was cleaned and made suitable for loading. The whole incident was a reminder that some fellows you meet in the business are awful, some are outlandish, and some are simply dense.

PART SIX

DIRECTIONS, SHIPPERS, STRIKES, AND BABY ANIMALS

DIRECTIONS

Understanding directions can be as challenging for some drivers as taking exams are for some students, and I've certainly met my fair share of drivers who were missing pallets in this department. One time, after Joe, a WMTS truck driver, made a delivery of building supplies, he called to tell me he was empty. I asked where he was and heard him ask someone for the name of the establishment where he'd delivered. He then came back on the line and said, "I'm at ABC Supply."

"Okay. I meant, what *town* are you in, Joe?" I asked. His delivery was only 170 miles or so from Baltimore, so I couldn't imagine he was too off track.

"Hold on a minute." There was a pause, and I'm sure he had to ask again. Then, "I'm in Little Creek."

"Damn, Joe, I've never heard of that one. What state?"

"Hold on a minute."

I heard him ask what state he was in. I was incredulous, but felt sorry for him when I heard uproarious laughter in the background. I suppose ignorance truly is bliss, because Joe came back on the line, and didn't seem to have a clue-in-hell that those folks were laughing at him. He told me he was in New Jersey.

"Joe, do you know where we're located in Baltimore? If you do, then just come on down here."

He said he'd try, and I hung up. I wasn't about to stay on the phone and listen to him ask someone for directions to Baltimore.

On another occasion, a driver who was also likely missing his last two pallets, or more, was dispatched to pick up a load of steel on the west side of Baltimore. He should have taken an exit off I-695 to reach the shipper, but for some reason or another, he missed his exit number. Instead, he kept driving, thinking he would eventually find the exit. He drove the entire circumference of I-695, over fifty miles, looking for his exit. Then, somehow, he missed his exit again and completed yet another revolution of Baltimore.

The driver must have thought that his third time would be the charm, because he went around Baltimore one more time. Finally, he decided he should stop and find a pay phone. After driving damned near two hundred miles, plus paying three separate tolls for the Key Bridge, he eventually found the shipper. It was a damned good thing that he had been dispatched early that morning, or he would have missed his pickup.

It's hard to imagine a world without cell phones, but back in the day, truck drivers had to find pay phones in order to perform their jobs. Every driver had to perform a check call each morning to inform their dispatcher if they were unloaded, loaded, or en route to a delivery, including an approximate time they would arrive. In the event of a truck breakdown, or a flat tire, the driver had to once again find a pay phone in order to get help. Often there were pay phones located at gas stations, so the driver would have to exit a highway, if it was an interstate highway, and hope that there would be a gas station at that exit. If there wasn't, he'd have to get back on the interstate and drive to the next exit to try again. If a driver had a flat tire on one of his front steering tires, he couldn't "limp" into the next highway exit. If this happened, drivers locked their truck cabs and stuck their thumbs out to hitchhike a ride to the nearest pay phone. Local calls cost five cents in the good old days, and came to cost twenty-five or fifty cents toward the end of their popularity, or the driver would call collect and have the receiver pay for it. If you talked too long, the cost increased. And, of course, there were no electronic devices giving drivers directions to shippers or receivers, and because a map could usually only get them to the middle of town, they usually had to find a pay phone in their destination's town or city to call the consignee. The drivers would have to look up the telephone number of the business by utilizing a phone book, which was always attached to the phone booth with a small, sturdy chain.

After writing down the directions, using paper and pen they carried with them, off they would drive to their destinations. If the person they had spoken with had given them the wrong directions, it became necessary to find another telephone and call back.

Drivers also often had to continually access phone booths in order to call their dispatcher. Say a North Carolina-based WMTS driver unloaded in New York and needed to know where to go next. He'd call his dispatch office to inquire where to pick up his next load, but if no load was available, the driver would be asked to call back periodically until a load was found for him. If the driver was told to begin heading toward the Baltimore terminal and call along the way, he had to locate an available pay phone each time he needed to call. This was the modus operandi prior to CB radios and cell phones.

The whole process was always a pain in the driver's rear end. Many drivers would exit a main highway in search of a pay phone and have to manage endless obstacles, like narrow streets, telephone poles, street signs, guardrails, fire hydrants, one-way roads, and cars parked everywhere, many of which were inadvertently destroyed as large tractor trailers tried to maneuver around them. More than a few gas station awnings were also knocked down because of a thirteen-and-a-half-foot-tall trailer trying to fit under an awning measuring twelve feet high.

Buster, one of WMTS's drivers with a great, dry sense of humor, called his dispatcher one morning to inform him that he was unloaded and was awaiting a dispatch for his next load. The dispatcher informed him that nothing was available at the moment and asked him to keep calling back. That afternoon, Buster called every hour from the morning to the afternoon, until he finally said, "Well, goddammit, I have wasted a lot of my expense money on pay phones, calling you all goddamn day long, are you going to give me a fucking load or should I just pitch a fucking pup tent for the night?" We all understood his frustration.

Today, because of a truck's onboard communications and GPS, drivers often don't need to call a dispatcher to know where to pick up a load. But this also means we can track trucks, in a way we couldn't before. Back in the day, if drivers wanted to goof off some, they could easily do so because no one knew where the hell they were. Thanks to the GPS units on top of trucks, truckers cannot hide these days.

I was operations manager for a company at the time it began installing satellite communications in all its tractors. Several months after installation was complete, a driver called our dispatch office to report that he was running a little behind on his delivery to New Jersey, but said he wasn't far and would be unloaded before lunch. He said he had just pulled into a truck stop north of the Delaware Memorial Bridge. I don't remember which truck stop, or which town he said he was in, but while we were talking, I pulled his truck up on my computer screen.

I said, "Just so I know exactly how long it will take before you arrive at your destination, what town are you in again?"

He gave me the town's name again, so I replied, "No, you're not. You're in Baltimore!"

"Hell no, I'm not! I'm in New Jersey," he declared.

I asked him if he hadn't noticed the new white gizmo on top his truck. This seemed to startle him and he said he'd seen it, but "didn't know what the fuck it was."

I told him it was a satellite tracking device, and I could pinpoint his vehicle within a few hundred yards. Later, back at the terminal, he wasn't bashful when admitting, "Man, shit! You sure did get the goods on me, didn't you? Y'all got all this new shit looking at us that we can't get away with nothing these days."

We soon learned that this one lie was just one of many he had in his repertoire, and his employment soon ended at our company.

SHIPPERS

Trucking folks know how badly some shippers treat truck drivers. Drivers were treated like shit years ago, and I'm not sure things have become a whole hell of a lot better through the years. One particular area of mistreatment concerns the weight that shippers load on trailers. Of course, all truckers understand the reason for shippers to load trailers up to the legal weight limits: because they don't want to risk sending a shipment that is smaller than it needs to be. (For example: the legal weight limit for a five-axle tractor trailer, which is the most common commercial vehicle plying US highways, is eighty thousand pounds and if a tractor trailer weighs in at seventy-seven thousand pounds, then the shipper has left three thousand pounds off the trailer that it cannot get paid for.) Some shippers initially load too much product to err on the high side rather than the low side, but too many of them push that far beyond what's necessary.

When shippers load heavy products, such as sheetrock, plywood, cement, ingots, or any number of other commodities, many times they require the driver, after securing his load, to weigh it, and then return to the loading dock with his weight ticket to receive his bills of lading. Some shippers have scales on site, but many times drivers have to travel to the closest certified scale. Every driver, whether he pulls a flatbed or van, also knows what might happen upon returning to the shipper after weighing his rig. If the weight is low and there is room left on the trailer, more products will be loaded. If the load is too heavy, they'll remove some of the product—and hurrah!—have the joy of returning to weigh the load one more time.

In the past, most shippers gave little or no consideration to the countless hours that drivers wasted going back and forth to weigh their loads, since it was thought to be part of a driver's job. Many a driver has weighed his load and found it to be either over gross weight or overweight on an axle. Quite a few drivers thought, *To hell with it*, and took their chances by running around, or bypassing the scales at state weigh stations using back roads. There were some trucking companies that would, if you got caught, pay your overweight fines. However, for the companies to pay, the driver usually had to first get his dispatcher's authorization to "run the load." If the driver decided to do this on his own, and then got caught, he would often have to pay the fine himself.

Through the years, there have been several high-profile crashes involving overweight trucks. Ensuing lawsuits made shippers painfully aware that both they and the truckers could be held liable for knowingly overloading trucks. Because of these rulings, shippers today are far less likely to load a trailer over the allowable gross weight for a five-axle vehicle (eighty thousand pounds), although axle weights still need to be adjusted pretty often. Federal law stipulates that the steering axle, the front axle, cannot exceed twelve thousand pounds, and that the other two sets of axles, the drive and tandem axles, cannot exceed thirty-four thousand pounds. If an axle exceeds the allowable weight, then, most assuredly, the driver is going to get ticketed for his transgression. It was high time that shippers learned that, "Not my problem," was now also their problem.

Most deliveries today are scheduled for a specific time slot, and the standard practice, unless other arrangements are agreed to, is for a receiver to be allowed two hours "free time" to unload the truck. If the receiver takes more than two hours, most trucking companies will assess a "detention charge" for holding up the driver and the tractor trailer. Large trucking companies are mostly successful in receiving detention charges after two hours, and most of their drivers are paid a portion of these amounts.

Owner-operators haul the vast majority of produce in the United States and, sadly, too many of these produce haulers don't receive detention fees; despite the fact that they often have to wait several hours for their turn to load in the farm fields, and have to apply the KY Jelly, bend over, and grab their ankles when they arrive at their destinations, because they know damned well that even though the load was hotter than the blazing gates of hell when it was loaded, the lumpers really don't give a shit how long it takes to unload the truck.

A lumper is a person who unloads cargo. In the not-so-distant past, truckers were required to unload their own loads, but today practically all food warehouses ban truckers from even stepping foot on the unloading docks. Instead, on delivery, each trucker hires a lumper to unload the product. Lumpers' rates depend on how much time they spend unloading a trailer. I have seen these charges range anywhere from $75.00 to $500.00 per load. It may sound like good money, but it's not an easy job. Just try unstacking and restacking forty-eight thousand pounds of product when it's ninety degrees outside and the sun is beating down on a metal roof.

After produce haulers arrive at their delivery destinations, they often have to wait several hours before receiving a dock assignment and then several more hours, typically four to six, while the lumpers do their thing, and the owner-operator waits for the unloading to be finished. And while most large trucking companies receive detention pay after two hours, owner-operators are hardly ever paid for their detention time. The main reason for this is because the owner-operators contract their loads through food brokers, and once the driver agrees to haul a load for a negotiated sum, then that is all the driver is going to get.

Everyone is aware that trucks are the only reasons buildings have loading docks. Trucks haul products in and out of businesses. Even a dunce could understand this, so why is it that so many shippers and consignees treat truckers like redheaded stepchildren? Truck drivers ask shippers if they can use restrooms or get something from a break room, only to be told that truck drivers are not allowed to use their bathrooms and that the break rooms are only for employees.

I have been that driver who has eaten too many apples and needed to use a restroom, and quickly. I have also worked in shipping and receiving facilities and observed my employers do these very things to truckers who were delivering to, or picking up from, our facilities. Is it any wonder that shippers ask questions such as, "Why do truck drivers have to take a leak alongside their trucks?" or "Why do drivers throw their trash, their banana peels and their chicken bones, out on the parking lot?" The shippers and receivers could get an earful of answers if they would just ask the truck drivers.

While on the subject of bad shippers, truck drivers often walk into shipping or receiving departments and are greeted by the sight of someone's back. It is as though those folks are thinking that if they ignore you, you might go away. When truckers get tired of standing there being ignored, they often clear their throats loudly several times. Eventually, without even turning to face the drivers, the shipper finally says that they will be with them in a minute. And the hell of it is, if truckers say anything at all about being ignored, especially if the trucker calls the shipper out on it, the driver may end up being the last truck unloaded that day or even get banned from going back to that location.

When there is a scheduled time for a pick-up or delivery, the process should work better, but with a bad shipper, this is far from guaranteed. There are numerous ignominious examples of truckers who have suffered despite having scheduled times. On many occasions, drivers run hard, and even forgo meals, in order to arrive at a customer's business before his federally mandated on-duty driving time[3] runs out, only to hear, "You're too early. Wait until thirty minutes before your appointment and then we'll let you through the gate," or "The conveyor line went down, and we will get to you when we can." A driver might also arrive tired as shit only to find an extremely long line of trucks ahead of him. He might think he'll be able to get in a nap while he waits, but then realizes that if he tries for some shut-eye, one of those other sons of bitches will drive around and get in front of his truck or, the company's dispatch office will call him on the CB when it's time for him to load and if he's asleep, he might not hear them.

You could write an entire book about bad shippers. Thankfully, you could write a sequel about the good ones who look at the truck driver as an essential component of its supply chain.

Unless they work in the transportation and logistics fields, most folks wouldn't give much thought to the job of a company's traffic manager (TM). Well, John Q. Public couldn't imagine the power of a large

---

3   The Federal Motor Carrier Safety Administration regulates the number of hours commercial vehicle drivers can operate, which, simply put, presently allows a trucker to drive no more than eleven hours before taking a ten-hour break.

company's TM, as he dictates all transportation decisions, whether inbound, outbound, global, or warehousing, nor could John Q. fathom how this power directly affects trucking companies.

Before deregulation led to lower freight rates, many carriers spent enormous amounts of money literally "buying" their freight and, sadly, there happened to be some TMs who were ripe for the picking. Most TMs were, and still are, honest and hardworking folks, however, there were some memorable fellows who thought that carriers should pay for their loads, and these TMs lived very high on the hog. I know of TMs who never, ever paid out of pocket for their own family's vacations, even though they took several trips each year. One TM drove a new Cadillac each year, compliments of a trucking company. It was astounding how much freight could be bought with a bottle of good bourbon or a case of tolerable wine.

Traffic managers negotiate—or dictate—freight rates with carriers. TMs often supervise large numbers of employees, including dispatchers and related clerical staff, and it used to seem that a TM's entire staff mimicked their boss's behavior when it came to freebies. The line of thinking went like this: If he was getting his, they were getting theirs.

Most steelmakers' transportation personnel were quite upset when carriers missed picking up steel loads. They were not in the least bit sympathetic concerning the reason(s) for these no-shows, and they would cuss you like a dog in addition to withholding a carrier's freight for a day or two. Shippers hate no-shows because when a load has not been picked up, the customer does not have its product, so the shipper loses revenue—and, yes, the carrier does too. Usually, there are good reasons for no-shows, such as driver sickness or equipment breaking down, but as a traffic manager once said, "I don't give a fuck if the driver died—a no-show is still a no-show."

A traffic department dispatcher called me one morning, just as the clock struck eight, and greeted me with, "Way to go, asshole, you missed a fucking load last night." He didn't even let me explain, he yelled and then hung up on me. Well, I think I gained a bit of his respect when I returned his call and told him that I didn't give a shit who he was, he had better never call me an asshole again. He never did, and we maintained a professional business relationship for several years after. I knew I wasn't the only one who shared this kind of beautiful moment with him, because soon after I started hauling loads he assigned, other carriers asked me if he'd cussed me out yet.

When I was running the WMTS Pittsburgh terminal, it was only fitting to take the TM of one of our biggest flatbed shippers, Art, out to lunch from time to time to maintain a good relationship. This TM controlled shipments from four or five plants that manufactured mining equipment, and also imported shipments from the Port of Baltimore. Those were the high-flying days of the three-martini lunches, and Art fondly embraced them. I would pick him up out-side his office at 11:30 a.m., and we would head downtown to one of the Tambellini restaurants. There were several Tambellini's around Pittsburgh, and they were all excellent. Art would order a vodka mar-tini with pearl onions, and I would enjoy a gin martini with olives. As well as I can remember, our conversations must have been very engaging, very lengthy, and, to be sure, quite debauched, because the end of our lunches always coincided with just barely enough time in the evening to pour him onto the last downtown bus to take him to his home in the North Hills.

At my first lunch meeting with Art, I kept up with the number of martinis he drank. The resulting hangover, possibly amplified by the richness of Tambellini's food, is one I've never forgotten. I had to behave myself at subsequent lunches with him, and switch to water after only two martinis. It made it a hell of a lot easier to drive home after dropping him at his bus stop.

I also took the dispatcher who called me an asshole to lunches and dinners, but thank God I never had to spend seven-and-a-half hours with him drinking martinis.

The TM at a large steel mill in Baltimore was tough as nails, but treated his employees and carriers fairly. Our company hauled a large volume of steel loads from all the different mills within this complex, including coiled steel, tinplate steel, and steel pipe. On the first evening I asked the TM to dinner, he requested Haussner's Restaurant, which was located in the Highlandtown area of Baltimore and was one of the best-known restaurants in the city. An entire wall of the eatery's second floor was painted with a mural of a battle that took place during the War of 1812. Ornate oil paintings that hung throughout the restaurant, including dozens of nudes in the bar, were worth millions of dollars.

At least once a month, this TM would call our office, usually a bit before 5:00 p.m., and ask whether I'd like to go to dinner. He knew that I could hardly turn him down, due to the volume of business he gave our trucking company, so I'd meet him in Haussner's between 5:30 and 6:00. After several cocktails, we'd finally order food. I really never knew what we ordered, but it always managed to be very expensive. And you can be sure that we washed those delicacies down with more cocktails, and bottles of good wine. One thing that sticks with me is that at the end of each meal, he'd order a strawberry pie to take home to his wife. Haussner's pies were rich, decadent, and expensive. But because the economy was in high gear at the time, my company never questioned my expense accounts after dinner with this TM.

On my way home after one of these dinners, I needed to pull over onto the shoulder, under an underpass on I-95 North, to rest a few minutes. When I opened my eyes after what felt like just a few minutes of rest, I looked at my watch and saw that it had somehow become six o'clock in the morning. To this day, I still wonder why I wasn't awakened by law enforcement. Since this was before we had cell phones, I had one hell of a time explaining to my pregnant wife why I was just coming home from a dinner that had begun at six o'clock the previous evening. She probably still thinks I was out with some floozie.

One transportation dispatcher, in his desire to imitate his TM's actions, took his greed to another level. Carriers referred to him as a "whore" because of the way he acted at dinners. If you took him to the most expensive restaurants and watched as he ordered two of the most expensive entrees, then your company could expect a large volume of shipments. He would order two filet mignons, or two whole lobsters, or two whatever-the-highest-priced entrees happened to be on the menu that evening, and he would request one of the entrees to boxed to-go after being prepared. He'd also order two desserts—one to go. It was a bit ridiculous, but he was a likeable fellow who was so close to retiring in Florida that he didn't give a hoot what people thought of him. All carriers, including my company, wondered if he would be able to buy two of everything when someone else wasn't paying the tab.

Another traffic manager wasn't the least bit bashful about calling to inform me that he was closing in on the last bottles of the excellent case of wine I gave him on a regular basis. One time I called to ask why our level of steel shipments had been slow for several days, and he replied, "Oh, I thought I'd told you I was out of wine." I suppose it was necessary to grease the palms in order to grease the wheels.

STRIKES

The fact that I grew up in Western North Carolina, with a father and grandfather who ran things themselves, practically ensured that I would never be a union man. My college experiences never caused me to lean left of center on the political spectrum, and to this day, I think a person is better served when others do not speak for him or her. I was born in a right-to-work state, and although there seem to be several different definitions of the phrase, I define it as having the right to work without some union group trying to prevent me from doing my job. Also, it has always raised the hair on the back of my neck when I've been told that I cannot do my job, or make money, because other people are striking. That's my personal feeling—but I do respect the thousands of truckers who prefer the union way. We have a kinship in being truckers, which stays with us even if we have ideological differences.

The Fraternal Association of Steel Haulers (FASH), comprised of steel-hauling owner-operators, was founded in the late 1960s, as an offshoot of the Teamsters Union, which had previously represented most of its members. FASH's founders thought the Teamsters overlooked the needs of the owner-operators, and that they deserved better representation. They wanted detention pay for the lost work hours, or days, they spent in steel mills waiting to be loaded, but the Teamsters had ignored their pleas for help securing this. In the late 1960s, it seemed as if Teamster President Jimmy Hoffa had adopted a policy of not caring about the needs of owner-operators.

I understand why FASH was pissed off by this, but I don't stand behind how they tried to achieve their goals. For example, while I wasn't their enemy, their protests caused me to endure several terrifying weeks of driving. During my breaks from college, I suffered at their hands during several unpleasant dispatches that caused me to traverse the Pennsylvania Turnpike from Breezewood to Pittsburgh and beyond. They were terrifying journeys because the FASH boys had decided that while they were on strike, no truck drivers should be on the road, as a form of solidarity, and any driver who was on the road deserved to be punished. Consequently, they would send bricks and cinderblocks toppling off the turnpike's overpasses and onto passing trucks.

Since I wasn't going to not make my runs, I had thought up all kinds of ways to protect myself from their window-breaking missiles. My main strategy was to make my truck a difficult target, so I used every bit of the roadway as I arrived at each overpass, moving my truck between lanes and both shoulders, and even moving into a different lane as I went under each overpass. I would also cover my eyes with one of my arms as I went through the overpasses. I knew that I couldn't block the entire windshield, but I thought that I might be able to at least protect my eyes if I was hit. This wasn't easy, or particularly safe, to do while passing, or being passed by, another truck, or when there was heavy traffic, but I did it when I could. During these episodes, I was by far the most scared shitless I've ever been while driving. I offered up many *Thank you, God*s when I drove this turnpike.

The strike lasted two weeks, and while drivers don't usually like it when law enforcement officers hang out on overpasses, during this time, every trucker on the road was grateful when the cops were out.

The next time there was a FASH strike was when I was a terminal manager in Pittsburgh, and one of our trucks was headed to deliver a closed-van load to a consignee near the Pittsburgh airport. When the driver stopped at a red light several miles from the consignee, a fellow climbed onto the steps of his International tractor and asked the driver if he knew there was a strike going on. The driver replied that he was aware that FASH drivers were striking, but he wasn't on strike, so he needed to deliver his load. The FASH member told the driver not to come back to that part of town again or there'd be trouble, and that he would be smart to park his truck.

While the striker was standing on the side of the tractor, he purposely blocked the driver's view of his side mirror, so the driver did not see another striker reach under the trailer and pull the handle on the fifth wheel. The fifth wheel is what connects the tractor to the trailer. When the driver started moving after the light turned green, the trailer became unhooked from the tractor and slammed onto, and then off, the tractor tires, and then fell onto the landing gear pads, which were raised for travelling. The trailer blocked part of the intersection, and some Good Samaritan must have called the cops; but even though they arrived quickly, by then the FASH members had fled the scene and were nowhere to be found.

The trailer's landing gears were bent and could not be hand-cranked, so the police summoned a wrecker to lift the trailer in order for the tractor to get back under it. Thankfully, no one was injured, and no other vehicles were involved. The driver was an older gentleman, and handled the situation like it was just part of his job. Those FASH bastards, however, cost our company quite a bit of money from this. I wonder if they would have been proud of themselves if the trailer had split from the tractor, built up speed, and then run into a busload of kids.

During another one of my breaks from college, I arrived at a Greensboro steel distribution facility with a load of steel coils at around eight o'clock in the morning. Our dispatch office had alerted me that the plant workers might be on strike, so I wasn't all that surprised when I saw the entrance gate blocked by fifteen to twenty striking steelworkers. They refused to move and let me through the gate and I quickly learned that I would forever be labeled a "scab" if I went into the plant. But I still had to make my delivery, so I asked the gate guard to contact someone in the receiving office. Soon, a supervisor came out to let me know he'd called the Greensboro Police Department and requested that they send officers to help me get into the plant. Several police cars arrived soon after, and the officers told the strikers to move out of the way. There were loud, obscenity-laced protestations, but eventually they parted and I drove through the gate and into the plant.

The supervisors and foremen were more helpful than any of the steelworkers had ever been, and several of them climbed onto the flatbed and helped me remove my tarps and chains. It took less than thirty minutes for the shipment to be completely unloaded. They also helped me fold my tarps. It was great.

When I returned to my truck after washing my hands, one of the supervisors told me that picketers had gone around the building and were now blocking the roll-up exit door. He didn't know how I would get out. I judged the distance from where I was parked to the exit door, and guessed that I could be in fourth or fifth gear by the time I got there, so I asked the foreman to raise the door when I started moving. Sure enough, the door opened to reveal the strikers defiantly standing there, but I was headed toward them at a fairly good clip. Of course, I still had room to stop if they didn't get out of the way—I would never intentionally run someone over—but I didn't even have to slow down because they scattered when they saw me coming. It probably helped that I was driving a truck that had an extremely loud 903 Cummins engine.

I am no badass, and under any normal situation, I wouldn't drive at people knowing it would scare the heck out of them. The only reason I crossed that picket line was that I was being prevented from doing my job. After I passed the strikers, I didn't stop at the stop sign at the end of the street or make a complete stop at any stoplight on my way to I-40. I also didn't let many cars stay alongside me very long as I drove east. I didn't want to get shot at. I don't regret having driven through the picket line, but I'm glad I never had to do so again.

BABY ANIMALS

During the middle of a hellacious snowstorm one winter, a scroungy, malnourished, dirty white dog staggered into the Baltimore trucking yard and tried to shield itself from the snow by lying down and taking cover between the two fuel pumps. Each time Frank and I tried to approach her, she ran from us. Eventually, after we placed a bowl of warm milk beside a fuel pump, she realized we could be trusted. Since she had no collar or tags, we couldn't bring her back to her owner, even if she had one, so we decided to take her in and give her a new home.

Frank named her Baby and she quickly became his assistant while he worked as a mechanic, and his best friend. A veterinarian gave her shots and deworming pills, as a precaution, and informed Frank that she was in good shape. Good dog food quickly took care of her malnourishment.

For the next two years, if you saw Frank, then you saw Baby. If Frank was working underneath a truck, Baby was lying beside him. If he was operating the forklift, she was sitting behind him. If he went on a road call, she was sitting beside him in his pickup truck. When he walked out on the yard, she was by his side. But she also knew when to stay out of the way, lying on her blanket when equipment was moved in or out of the shop.

One horrible afternoon, Baby inexplicably left her blanket and began walking across the shop floor just as Frank was backing a tractor out of the shop, and one of the tractor tires hit her. Horrified, Frank scooped her up and rushed her to the vet. The vet said she had some broken bones in her hip, and while she might be left with a limp, she would otherwise recover. Afterward, Baby did a lot of lying around and sleeping, but her appetite was good, so it seemed like she was on the mend.

A few weeks later, Frank got to work one morning, and Baby failed to greet him. When he turned the lights on in the mechanic's bay, he found her lying on her blanket. She had died during the night. In the autopsy, Baby's vet found that her stomach contained enough antifreeze to kill her. Frank figured that Baby had smelled the sweetness of an open container of antifreeze, and, who the hell knows why, drank it. Frank cried like a baby for his Baby, and I know he always blamed himself for leaving the antifreeze uncovered and out in the open. He never had another dog at the terminal.

I was no longer working at the terminal when Frank lost Baby, but word of it spread quickly. Years later, when I was in our first Pittsburgh terminal, I found my own best friend while at work. At the time, we were renting the terminal from another trucking company, which used the offices on the top floor of the building, and we had the entire bottom floor. The other company's owner and its dispatcher, Debbie, were seasoned trucking folks, very nice people, and we all got along famously.

A couple of months after we moved in, I came to work one morning and found a little white ball of fur leashed to a fence post alongside the building. She was barking and jumping around, and when I began to pet her, she was so happy that she peed on the ground. She was sweet and playful, but I could tell that she had feistiness to her too, and while she didn't bite me, I could see that her little teeth were as sharp as razors. Turns out that someone had gifted Debbie the ten-week-old puppy, a mix between a German Shepherd and an American Eskimo, and Debbie had named her Sheba. Debbie was unable to keep Sheba at her house, so she decided to leave her tied up outside at the terminal.

I have never kept a dog chained up, and I wasn't all that comfortable with it, so after I arrived at work each morning, I would unleash her and let her run around all day. Debbie thought this was great for the pup, and soon realized how well Sheba and I got along. After a few weeks of this, she asked if I wanted to keep Sheba, and so began a long and faithful friendship.

It didn't take long for me to learn that Sheba was pretty damned smart. At four or five months old, I decided to begin the process of teaching her to sit up. I quietly spoke to her with the calm instruction to "sit up," as I pushed her rear down into the sitting position, and lifted and held her legs up. At first, when I let go of her legs, she toppled backward as if she was in slow motion. But then she got up, licked my face, and sat down in front of me. I raised her legs, released them, and she sat there! She had it down pat after one try. Through the years, she would often drop balls or shoes at my feet, and then sit up, while waiting for me to throw them. If I happened to be on the phone, she would remain sitting up for fifteen to twenty minutes. I eventually became accustomed to her just sitting there, but observers could not believe the length of time she would sit up and patiently wait for me to throw the ball.

Sheba went to work with me at three different terminals, practically every day for seven years. She was always outgoing and happy to meet new truck drivers. In Baltimore, her favorite game was to drop a tennis ball at someone's feet, back up a step or two, and then try to prevent the ball from getting kicked past her. After mowing my lawn one weekend, I posed Sheba so her paws were on the mower's push handle, and then snapped some pictures from different angles. When I shared the photos at work, most everyone looked at the pictures, and whether they were office personnel, mechanics, or truck drivers, just about everyone had the same comment, "Yeah, right." Later though, in the middle of the afternoon, one of the mechanics walked into the office, and, noticing we were alone, walked over to me, and quietly asked, "Does that dog really mow your lawn?" I explained to Don that although Sheba required some assistance going up and down the hills, she always did a great job mowing the level parts of the yard. As he shook his head back and forth, he declared, "Goddamn, that is one smart dog!"

Sheba had quite the impressive trucking resume, so it only followed that she'd need a headshot to boot. One day I sat her in an office chair at one of the desks, inserted a telephone receiver into her collar, and then snapped several pictures. She made a great a dispatcher and was also a hell of a lot more honest than Slick Jack.

Sheba was with me when WMTS transferred my family and me from Pittsburgh to Baltimore and when we moved back to Pittsburgh a year and a half later. By the time I returned, the Pittsburgh terminal had moved, and the new office was leased from a different trucking company. They had two employees, whom I got along with well. They had been in trucking for many years and had a great sense of humor, and they were glad that Sheba came to work with me every day. When none of my drivers were around to play with Sheba, one of those fellows would play ball with her in the hallway.

I happened to be in their office one morning when the terminal manager drew a map and directions for a driver onto a piece of paper. With the map in hand, the driver left to pick up a load of steel some thirty-five miles north of our terminal. For some reason, probably to check on Sheba, I had gone back over to the carrier's office, and had to wait until the terminal manager finished his call. The driver was calling to say that his load had been cancelled, and asked what he was supposed to do. Since there wasn't another load close enough for him to pick up, the manager said he should just return to the terminal. The poor fellow must have been missing those last two pallets, because he asked the manager how to get back. The terminal manager, quite irritated, answered, "You know those directions I drew for you? Turn them upside down!" With that, he slammed down the phone and made numerous disparaging remarks about the driver's mental aptitude, and the driver's mother. He then looked down, noticed Sheba, and pronounced, "Sheba, you have a hell of a lot more sense than that son of a bitch I was just talking with."

The truck parking lot at this terminal must have been at least ten acres, which left me plenty of room to hit golf balls. It didn't matter how far I hit the balls, or where they went, because Sheba located and returned each one. The yard was also great for hitting tennis balls to Sheba with a baseball bat. Inside, there was a long, four-foot wide hallway separating the two offices, which provided an excellent venue for Sheba to play her game, "See if you can get the ball past me." There was one driver who especially enjoyed playing her game, and did so each time he came to the terminal. He played with her for a long time one afternoon while he was waiting for his loading appointment at a steel mill. When it was time for him to leave, he stuck his head into my doorway and declared, "That dog is a four-legged Willie Mays!" This was Sheba's best-fitting description and one of her finest compliments.

I was working out of town when my wife called to inform

me that Sheba was "down." She couldn't stand up, eat, or drink. Circumstances made it impossible for me to return home for a few days, so my wife and I agreed that she would take Sheba to the vet for him to put her to sleep. She lived just one month shy of eighteen years. I had never thought of what to do when she passed away, but her ashes still rest in our family room.

Most everyone who's owned a dog thinks they've owned the smartest and most well-trained dog. Just mention how smart your dog is and you will soon learn all the amazing things other folks' dogs can do—we're talking about *smart* animals here, not cats. We love our pets. We take them with us when we go for a drive. We take them with us when we go on vacation. We act silly over our pets—doting on them, spoiling them, hell, even dressing them up—because they are part of our family, our best friends. Did you hear the one about the guy who asked, "Why are dogs better than a wife?" His answer was, "Because your wife gives you hell when you come home at 2:00 a.m., but your dog licks your face and is terribly excited to see you!"

Animals bring joy, laughter, and comfort to a great many people. I met Sheba as an energetic ten-week-old pup and the gal grew into a beautiful, pure white, fifty-pound best friend. She was so special to me that I cried my eyes out when she died. Over twenty years later, I still think of the things she used to do.

I've also seen a few cats find their way to the trucking yard over the years. When I was still driving for my dad, a black-and-white mixed breed cat found our furniture warehouse, and whether she found some mice, or more likely due to the fact that we started feeding her, she decided to stay close for a while. When the weather turned cooler, she must have enjoyed the warmer climate of the local pick-up tractor, which was a big, red, gas-guzzling, two-axle Chevrolet. We aren't sure where she came from, but she was very friendly and always came meowing when we arrived to start work.

One time, I returned from picking up some LTL (less than truckload) furniture at a plant in Shelby, North Carolina, and when I shut off the motor, I heard a soft commotion under the tractor's bench seat. Damn! That cat had somehow got into the tractor and delivered four or five kittens, and every one of them was crying for their mama. I gathered them up and put them in a small box lined with some clean rags. A few minutes later, I received a call from someone at the furniture plant I had previously left, who told me there had not been another driver at the location since I was there, so did I know anything about a cat which was running around their warehouse? Well, I loaded those kitties in my car and away we headed the fifty miles back to Shelby to get their mama. Once there, I parked the car, picked up the box, and started for the shipping office, but hadn't even made it to the door before the mama cat came running to the sound of her babies.

The kittens didn't cry at all on their way back home. They just ate and slept. Afterward, mama cat and her kittens stayed around the warehouse for six to eight weeks, but when we went to work one morning, they were all gone. I supposed the mama cat had just wanted a good, warm shelter in which to deliver her kittens, and then they found another home.

PART SEVEN

KNIGHTS OF THE HIGHWAY

In years past, truckers have had the distinction of being known as the Knights of the Highway, because they could be counted on to stop and help stranded four-wheelers. Through the years, these knights changed thousands of tires, extinguished numerous camper tire fires, added water to countless dry radiators, pushed vehicles out of the snow, saved many lives by pulling motorists from burning vehicles, and even delivered babies. Truck drivers have never expected to be paid for helping, and most refused offers of payment. They stopped to help because it was the right thing to do.

These days, some truckers still stop to help, and all stop in the case of emergencies, but the practice seems to have diminished. I suppose there are some obvious reasons for not stopping, such as the threat of being mugged or having your vehicle jacked. For these same reasons, most car drivers don't stand outside their vehicles and wave down help. They stay put in their locked cars and use their cell phones to call for help. High interstate speeds also make it impractical for trucks to stop because by the time truck drivers notice a breakdown, they are too far past the breakdown site to stop.

A Knight of the Highway once saved me from a costly tow-bill, not to mention possible admonishment from Pennsylvania law enforcement. It happened around eleven o'clock on a cold, snowy evening in Pittsburgh, when I was on my way home from a business dinner (one that either started late or, more likely, ran late because we shared a few after-dinner cocktails in the restaurant's bar). I'd left downtown through the Fort Pitt Tunnel and was headed up Greentree Hill. Although there were only a few inches of falling snow, plows had pushed it out of the road and into berms on the shoulders.

I thought I was driving carefully, but my Datsun (now Nissan) 240Z 2+2 somehow straddled one of the two-feet high snow berms. When the car came to a stop, I put it in reverse and let out the clutch, but absolutely nothing happened. I got out of the car and saw that I had perfectly balanced it on top the berm, and all four wheels were off the ground.

I was not dressed properly for slogging several miles through the snowy slush in order to find a pay phone, so I used my CB radio to beg some trucker to come to my aid. Using the trucker's channel 19, I broadcast, "Are any of you eighteen-wheelers heading up Greentree Hill? Could someone with a chain come help me?" I continued repeating the transmission as though it was my SOS, or Mayday call, throwing in things like, "Come on guys, have a heart," or "Help another trucker and pull me out of this snow."

Twenty minutes after I began my plea, a Knight of the Highway, in the form of an owner-operator, pulled up behind my perfectly balanced vehicle. He'd noticed me while he was heading up Greentree

Hill and then heard me begging for help on the CB radio. He said he'd chuckled a couple of times while listening to my broadcasts.

I knew this section of highway very well, so I was aware that this fine fellow had to have driven several miles, and upon reaching an exit at the top of the hill, done a flip and driven back down the hill to come help me. He then had to go around two different cloverleafs, so he could get back to me. With the power of his chain, it took all of ten seconds to pull my car off the berm.

I tried my damnedest to pay him for his trouble. I asked him for his name and address, so I could send him a check. When that didn't work, I asked if I could buy him breakfast. He declined it all and calmly explained that it just looked like I needed help, and he was glad to be able to assist.

Personally, I think another reason that truckers don't stop to help nowadays, especially younger drivers, is because they are unaware of their former status as Knights of the Highway, and some think like the Baltimore dispatcher who said, "No baby, that ain't my de-PART-ment!" Perhaps if they understood the storied past of their forbearers, and the honor that inheres being one of the last great American cowboys, younger drivers would be more inclined to take on the role. Maybe truck driving schools should teach a course titled Knights of the Highway 101.

It's also true that the revised federally mandated hours of service limit drivers being able to "waste time" by stopping to help, contributing to the decline in the number of truckers observed pulling over to help four-wheelers. Some people may think I'm full of shit for saying this, but quite candidly, I do feel that truckers need the better image that could be brought about by helping others more often. The sight of a big rig helping a four-wheeler is a very powerful message to the motoring public. It's not that there aren't truckers who'll help, it's just that there aren't enough.

One thing, though, has never changed. The Knights of the Highway can *always* be counted on to offer their help to a woman who's standing on the shoulder alongside her car. Especially in hot weather, and especially if the woman is partial to short shorts and halter tops. These breakdowns have even been known to cause traffic problems due to the high number of tractor trailers pulling over to help just one young lady.

A NOTE ON SAFETY

Trucking is one of the most hazardous professions. Many drivers have been injured or killed while conscientiously performing the multitude of tasks required of truck drivers. Most drivers remember climbing up onto the catwalk to hook their air hoses and pigtail for trailer lights, only to slip and bust their asses on the wet, or snow-covered, or iced-over diamond plate steel. And what about the joy of opening the right trailer door, just before backing into the dock, only to have product fall on their heads? We have all suffered the pain, and even worse, the embarrassment, of falling out of a trailer because we lost our footing on a wet floor. Even Obie had his troubles—though he said that he never fell, the pavement always jumped up and hit him.

Numerous drivers, myself included, have also been cut by the sharp edge of the end of the Signode strap found on thousands of products, which keep shipment boxes tightly secured. These injuries used to be even more frequent because some receivers required truck drivers to cut the steel bands of each bundle. Those suckers acted like they had homing devices, because the taut straps would take off upon being cut. I have a scar on my left arm to show for it.

Flatbed and lowboy trucking offers truck drivers even more chances than closed vans to bust their asses. It just stands to reason that if you climb onto a trailer often enough, you will eventually fall. Until the past few years, there wasn't anything for a driver to hold onto, much less ladders to use. One way to climb up was by placing your fingers in the tie-down pockets at the rear of the trailer—which were actually steel holes used to insert steel poles for hauling pipes, but also used to insert wooden extensions on trailer side boards, similar to the side boards I used to keep the Vietnamese kids from stealing beer—and then stepping onto the bumper guard required by the Interstate Commerce Commission (ICC), and pulling yourself up. (In 1953, ICC required all trailers to have rear bumper guards to prevent cars from running underneath a trailer.) Another method, if you were young and agile, was to stand on the tractor's catwalk and hold onto the headboard as you glided around it. I used to be young and agile, but one time my hands slipped, and I busted my ass anyway while trying this maneuver. I did, however, once see a chiseled, muscular, young truck driver, whose physique would have allowed him to pose as the mythical Atlas, stand flat-footed beside a flatbed trailer, squat down, and jump the fifty-five inches up onto the trailer. It was a sight to behold.

Working with chains, binders, and nylon straps provides even more opportunities for injuries. If you're using a binder pipe to tighten your chains and a chain breaks, you almost always fall. You get up, most of the time, but sometimes you require a doctor's immediate attention. Many drivers have had their teeth knocked out when a binder has snapped open and hit them in the face. Quite a few drivers brag scarred faces, thanks to all the stitches that the binder wounds require. I know some drivers who consider their scars to be badges of honor, or even "badges of courage," and the scars provide what they claim to be visual evidence of the experiences they recount in tall tales about how they got injured.

Pulling older trailers could also knock a driver's teeth out. At times, when a driver was pulling an old trailer, he would be cranking a handle whose bolts and nuts had rusted from years of exposure to rain, snow, dirt, and road salt, and one of the rusted bolts on the handle or a rod would break. The result was hardly ever pretty if the handle snapped and hit his face: teeth gone, face cut, and sometimes, if he'd been using great effort, arm or back sprained, and knee cap busted open upon hitting the ground.

Working around equipment is also always hazardous, and most of us have the scars to prove it. I suffer a ringing in my ears, tinnitus, as many other guys do, developed from regular exposure to extremely loud noises, whether from driving very loud trucks, operating heavy equipment, listening to Obie beating the hell out of something on an anvil, dropping dock plates onto concrete, or forcing together three-piece rims using a sledgehammer, or a combination of all of these. Truckers today are damn sure to make our kids and grandkids wear eye and ear protection. I would be a hell of a lot better off if I had known to use ear protection.

But not all drivers' injuries are caused by the inherent dangers associated with trucking. Some, I'm sad to say, are due to a driver's desire to show off his agility and strength, and I admit I've made some brash choices myself.

One memorable time occurred when I worked in the office at WMTS. From time to time I was required to drive if we did not have a road driver available. On one occasion, the trip was to pick up a load of steel at Bethlehem Steel in Sparrows Point, Maryland, which no longer exists. So I pulled on my overalls, pre-tripped a tractor and flatbed, and then drove about eight miles to the facility, and backed into Bethlehem's Dock 48. I set up the trailer for a load of steel coils, using 4x4s and coil racks. Coil racks are twenty-four-inch-long pieces of two-inch-wide steel strips with four-inch, ninety degree steel angles welded onto the ends of the strips. The racks are laid perpendicular on the middle of the trailer and two eight-foot-long 4x4s are laid inside the racks. The racks provide a cradle in which to set steel coils.

The steel mill prohibited drivers from standing on their trailers, or sitting in their tractors, while the trailer was being crane-loaded, so I moseyed off to the men's room to kill some time. When I returned, I retrieved something from my tractor, and noticed that the trailer was already loaded. I closed the tractor's passenger-side door, and looked down the side of the trailer at the loading dock. The dock was the perfect height for me to showcase my physical abilities by jumping up on it, so I lit out running alongside my trailer. Using my last step to push up onto the dock, I saw myself as Superman: *Leaping tall buildings in a single bound.*

Goddamn, the pain was excruciating. My push step did not go as planned, as my foot planted in the middle of a puddle of steel coil oil. I'd only made it about halfway through the jump when my leg crashed into the steel-covered corner of the concrete dock. The material from my blue jeans and my blue overalls was driven into my dented shinbone and skin.

The steelworkers and truck drivers in the vicinity all witnessed my acrobatics. Even though my leg hurt so badly that I wanted to cry, I got up, brushed myself off, and walked onto my trailer as if nothing had happened. I felt as though I would pass out at any moment as I began the unbelievably painful task of chaining and tarping the coils.

I did not have stitches because there was not enough skin on my shin to stitch. Actually, what I really needed was a bone graft. To this day, I quietly wear both my badge of stupidity and my badge of embarrassment—it's Wrangler blue and fixed on my right shin.

I don't drive trucks anymore, but I still think like a truck driver when I drive my car. And given that I became intoxicated by the smell of diesel fuel at a very young age, it may come as no surprise to know that I now drive a diesel-powered school bus. Every time another four-wheeler acts in a way I consider stupid, or irresponsible, I try my very best not to lose my temper. *Try* being the operative word. But since I can do absolutely nothing about that bad driver, why should I get upset? While it's clear to me that many car drivers should be riding in taxis rather than driving themselves, the reality is that these drivers typically don't have a clue-in-hell that they are driving improperly. They cannot drive worth a damn and yet, in most cases, are totally unaware of it. But if *they're* not upset about their stupidity on the road, why should *I* get that way? I have a much more enjoyable drive when I don't get worked up every time a four-wheeler does something stupid. And clueless four-wheelers don't just piss off truckers—they are equal opportunity pisser-offers. They do it to the rest of us four-wheelers just as thoughtlessly as they do it to the truckers.

It's also never a good idea for a trucker to terrorize a four-wheeler, whether we like them or not. As a veteran truck driver, it embarrasses the hell out of me when I see truckers ride a four-wheeler's bumper while traveling seventy miles per hour. It embarrasses me even more to see the trucker travel in an unauthorized lane while they ride someone's bumper. I call this "arrogant driving." These truckers seem to think that they're somehow entitled to drive illegally in lane one or two of a four-lane road even though highway signs read: NO TRUCKS IN LEFT TWO LANES.

To America's truck drivers, let's return to the days when big rigs don't tailgate four-wheelers, and the motoring public isn't scared to be on the same highway as truckers. Let's return to the days when those same car drivers once again refer to you as the true Knights of the Highway. I would like to advise that the next time you play the game of ride-that-four-wheeler's-bumper, you need to become aware that you—not that car driver—but you, and only you, are the professional in this match. Remember that the car driver is an amateur. Act like a professional truck driver and move back, and away, from his bumper. Pretend the car ahead of you is carrying your children, and back off!

And for safety purposes, please also don't follow another truck too closely. We've all seen or heard about incidents where two trucks or buses (usually belonging to the same company) have both gone off the highway because the driver in front made a horrific mistake and the driver behind had been mindlessly assuming that the guy in front of him was driving properly, and followed him, without realizing the mistake, until it was too late. Some large trucking companies enforce a half-mile, or even one-mile, minimum distance between each truck, but sometimes we drivers enjoy drafting another truck, so we end up driving precariously.

One time, I was on I-75 South, driving through Kentucky with a sister truck, and the other driver was following within just a few feet of the rear of my closed van. I kept glancing at the guy from my side-view mirror and the fact that he was on my ass suddenly started affecting my driving reactions. We were traveling around sixty-five or seventy miles per hour and had just driven onto a newly paved section of the highway when I heard a loud hissing sound. I figured it was due to the new pavement, but then my tractor started pulling hard to the right. My right steering tire was going flat, and the other driver was right on my ass.

My hands were so busy trying to control the steering wheel that I couldn't grab the CB radio to tell the other driver what was happening. But I really didn't want him to hit my trailer, so I did my best to warn him by blinking my marker lights and turning on my four-ways, and I gradually began braking—all while my tractor was drifting off the right shoulder of the highway, and I couldn't seem to do a damned thing about it.

Eventually, I guess he figured something was wrong because through my driver's side mirror, I finally caught sight of him slowing down. With this assurance, I braked as hard as I could without wrecking myself. Lady Luck must have been riding shotgun because thanks to the repaving, the guardrail had been removed and I was able to drive into a flat and level field. By the time my truck slowed down, its tire was completely flat and I had traveled over one hundred feet into the field. Somehow in that moment, I had the sense to realize how hard it was going to be for someone to change the tire while my truck was in the hayfield, so while the truck was still slowly rolling, I wrestled the steering wheel back and turned it to the left just enough that I was able to park the front axle of the tractor on the shoulder of the highway.

Afterward, the other driver took me to a nearby truck stop, and we went inside and had coffee while waiting for the tire man to get his stuff together. While sitting in our booth, we overheard two drivers talking in the booth behind me, and one of them said, "Did you see that goddamn truck back up the road a few miles? Wonder how he ever got into that position . . . ."

Well, I kept quiet and let him just keep wondering why I parked that way. I learned an extremely important lesson that afternoon, and not only did I keep my distance from other trucks in the future, I was also adamant that other truckers never stayed that close to my ass again.

Just a few years ago, I finished a seven-year run working at the Maryland Department of Transportation's Office of Freight Logistics. My position allowed me the opportunity to work with the federal and state agencies responsible for motor carrier regulations and safety, and with research organizations, trucking industry groups and associations, and many private trucking-related companies. The agencies and groups were involved in all aspects related to commercial vehicles, including safety, equipment, hours of service, and truck parking, and the lack of truck parking spaces was, and still is, a very high-priority issue for them. One of the most important projects our state motor carrier office worked on was identifying the overall truck parking availability in the United States. We also undertook a study in Maryland to assess the in-state availability and sufficiency of truck parking spaces.

Anyone who has ever driven through the nighttime hours, whether by car, truck, or motorcycle, would have to have been blind if they didn't see trucks parked, mostly illegally, on the highway shoulders, on- and off-ramps, and basically any wide spot on a major highway. If a trucker is falling asleep behind the wheel, he certainly needs to pull over, but he needs more legal spots to do so—they're simply too limited today. The shoulder is a terrible place for a trucker to park and the sad statistics show that numerous four-wheel drivers have crashed into them, and sadly, often lost their lives in the crash. In reaction to this, I've heard some members of the enforcement community make comments such as, "Well, the car shouldn't have been driving on the shoulder of the road."

Sometimes the four-wheelers hit the trucks because their intoxicated drivers veer off onto the shoulder. Other times, the rear-end crashes happen because of the "moth effect," named for the insects who are mindlessly drawn to any light source. This is when a four-wheeler mimics the behavior of a truck. Many four-wheel drivers respect the fact that most truckers really do know how to drive safely, so they follow trucks down the highway, especially during adverse weather conditions, by watching the rear lights on the back of the trailers, and following them. Unfortunately, because they're focused on following the truck's lights, drowsy four-wheel drivers often don't pay enough attention and don't realize when a truck has left the highway and parked on a shoulder. Instead, they follow the truck onto the shoulder and barrel into it.

I implore truck drivers to not park on the shoulders of highways but to go to the next exit and use one of the ramps. If I were driving, I would try to find an on-ramp, because even though a car might run into the back of your trailer, the chances are much higher that his lower speed will make it so the collision is not fatal. Begging truckers to use ramps might bother various state highway administrators, but people die when they're going seventy miles per hour and hit trucks parked on the shoulder—and people usually don't die when they're going the speed you'd use on an on-ramp and—at that speed—hit a truck parked on the shoulder.

To get truckers to avoid parking on the shoulder when they need to rest, there needs to be enough parking spaces for them to easily access. Truck parking shortages usually happen at night or during severe weather events, because too many trucks are vying for too few spaces. Some truck stop chains would love to build additional parking spaces, but local communities have been very successful in lobbying their elected officials to deny the truck stop's building permits, relying on the old standby NIMBY (not in my back yard) talking points, and citing noise, pollution, increased traffic, and the belief that truck stops are sinful lairs of vice and prostitution.

On the other hand, some truck stop chains do not want more parking, because they do not see parking as a revenue generator. They should be ashamed of this—truckers are their customers, and they ought to support them. They should also recognize that if they don't have enough parking, trucks entering to purchase the truck stop's low-priced fuel will have no place to park and be forced to get back onto the highway and go to a different truck rest stop.

But chin up, truckers, help is on the way. Numerous government agencies, research organizations, and private companies are working to find solutions for the truck parking shortage. Several forward-thinking states are adding parking spaces at some of their rest areas, and work is also underway for truckers to be able to find available parking spaces through personal, hands-free mobile electronic devices, or even reserve a spot at a low cost.

All drivers, truckers and four-wheelers alike, should reach out to their city's or state's elected officials by phone, email, or snail mail to let them know there is a shortage of available nighttime truck parking and that this creates precarious driving conditions for *all* motorists and *all* constituents. When enough elected folks hear about the gravity of a problem—and worry about their reelection—they may begin considering it, and eventually, work to get things done.

In South Carolina in 2009, thirty-five-year-old truck driver Jason Rivenburg was robbed and killed while he slept in his truck at a closed-down gas station. Before he went to sleep, he phoned his wife and told her that he'd parked in the abandoned lot because all the rest areas and truck stops were full. At the time of his death, Rivenburg had a young child and twins on the way. After the devastating crime, Jason's Law was drafted, and when it finally passed in late 2012, it was included in a transportation reauthorization bill titled the Moving Ahead for Progress in the 21st Century Act. The law provided more than six million dollars in federal funding toward the construction and restoration of safe roadside parking lots where truck drivers can rest. Its projects included:

- Creating new dedicated parking areas for commercial truck drivers

- Creating parking spots for CMVs (commercial motor vehicles) adjacent to truck stops and plazas
- Opening existing facilities to CMV parking
- Making capital improvements to public CMV facilities that are currently closed on a seasonal basis, so that they can operate year-round
- Making it easier to access existing parking facilities

Unfortunately, it took a horrific act of violence to spur the creation of this bill. We can thank Jason's wife, who worked with her New York State representative to honor her husband and make something good come out of the tragedy. Someday, maybe the big-box retailers, such as Target, Walmart, and Home Depot, can solve the truck parking shortage by allowing trucks to park in their empty parking lots during specific nighttime hours.

PHOTOS

Me at a convoy point in Vietnam.

A haul in Vietnam. As equipment operators, we
tried to go everywhere "fully loaded."

The International and Chevrolet tractors of one of the fine
trucking companies I had the pleasure of working for.

Me and a dog named Bubba standing beside a gas-guzzling
Chevrolet tractor. The kittens were born under its bench seat.

Sheba dispatching truckers on a busy day.

The photo that made Don, the mechanic, ask,
"Does that dog really mow your lawn?"

EPILOGUE

I began this book by writing about some of the ways that people trucked nearly sixty years ago, or at least how my family got the job done. The job required long hours and maintaining the trucks was hard work, but trucking provided money to live on for my grand-parents, parents, and us four kids. Trucking today, thank goodness, is quite different from what it was when I was growing up. Marvelous progress has been made, due in part to technological improvements in equipment and petroleum, which have helped the industry become a better steward of the environment. Some of us aren't spreading our used oil on dirt roads to keep the dust down anymore. And unless you observe an older Mack dump truck, you hardly ever notice black smoke coming from a tractor's exhaust stacks.

Drivers today ride very comfortably in powerful tractors as they pull trailers equipped with wind-efficient skirting and trailer-tails. Not too many years ago, drivers had to climb Town Hill, on I-70 in Pennsylvania, in third or fourth gear. Now, most trucks climb the mountain without ever gearing down. They also either don't get to cheat or don't have to worry about making mistakes while filling out their logbook, as they ride in tractors equipped with electronic logs. The tractors also have lane departure, forward, and side collision avoidance technologies. And cab interiors are very quiet, which means that drivers don't have to turn their CD or radio volume way up. And it's not necessary to deal with the pain of locating and stopping at phone booths, or the danger of this, and no one overhears you when you speak with your family.

When I began writing *A Trucker's Tale*, all I really had in mind was the notion that I wanted to put some of my funniest trucking memories on paper. After writing the first story, I thought of another, and then another. One story bled into the next. One stretch of highway led to the next. There were stops, digressions, accidents, wrong turns, early arrivals, late arrivals, and incidental destinations. How else to write about a life spent trucking? These tales were written in diesel, and some passed from one trucker's CB radio to the next.

To my fellow truckers, friends, and acquaintances: I have tried to listen as each of you expressed your truly uninhibited feelings over the years. Thank you for letting me remember your voices as I've attempted to respectfully commit these stories to the page. Doing so has most assuredly made me recall the joy and fortitude of your friendship. Even in the event that no one ever reads the book (except my family members whom I have threatened with bodily harm if they don't read it), my time spent writing has been worth the effort, as it's allowed me to relive so many good times.

ACKNOWLEDGMENTS

Never in my wildest imagination did I think I would ever write a book. So, imagine being confronted with the situation of thanking the many folks who have helped *A Trucker's Tale* become published.

To Lucinda Coulter, PhD, for advising, "Get it published!"

To Todd Dills for publishing parts of the book in *Overdrive* magazine.

To my fellow Navy veteran friend Les Taylor. If not for you, *A Trucker's Tale* would still be languishing within my computer. Your mastery of the English language is so damned enviable.

To my siblings, Betty Sue, Earl, and Yates, for a lifetime of memories. I am so thankful that we are as close today as we were as kids.

To my wife, Diana, for putting up with this trucker for all these years, and you actually like the book, too!

To my kids for suffering through the repeated telling of the book's stories over and over and over.

To my dad, Hugh, and my granddad, Obie, for teaching me how to become a trucker—and to cuss like one.

To my buddy Jeannie Fazio. You have always been my biggest supporter.

Thanks to the publishers and editors at Apollo Publishers, Julia Abramoff and Alex Merrill, for believing *A Trucker's Tale* was worth publishing. Julia has helped make this journey fun, educational, and sometimes frustrating, but never dull or boring. Her support has exceeded any I could have imagined. I am humbly honored to have Julia as my editor.

CPSIA information can be obtained
at www.ICGtesting.com
Printed in the USA
JSHW020248060523
41349JS00004B/5